Elizabeth Bernays has written a beautiful memoir. Her sense of amazement at nature and boundless curiosity make this account of the attention, inventiveness, global spirit, and fun of a life in science shine. Insects, under her devoted scrutiny, prove to be spellbinding theater—"a performance of great skill." This book offers a window into how and why to care about the smallest among us.
—Alison Hawthorne Deming, author of *Zoologies: On Animals and the Human Spirit*

From her childhood in Australia and schooling in England to the wilds of Hungary, India, and Africa, Elizabeth Bernays asks the reader to become her entomological lab assistant, to suffer the tropical heat and grime and army ant bites of a true field researcher. *Six Legs Walking* is a fascinating, beautifully descriptive, and lyrical narrative that captures the essence of some strange agricultural pests and the exotic places where they dwell.
—Ken Lamberton, author of *Wilderness and Razor Wire* and *Chasing Arizona*

Elizabeth Bernays's memoir illustrates what drives a life in science—persistent curiosity and a healthy dose of fearlessness. These serve her well in the face of obstacles encountered as she makes her way to success as a rare female researcher. Most enjoyable are her clear-sighted observations and original thoughts about living things and about the connections between insects and people she meets along the way.
—Nancy A. Moran, evolutionary biologist, 1997 MacArthur Fellow

I very much enjoyed reading these engagingly written memoirs of a great naturalist and superb entomologist. Her deep love for nature, the acute commitment to her pioneering research, and her warm sense for the people around her shine through on every page.
—Bert Hölldobler, Pulitzer Prize–winning coauthor of *The Ants*

Elizabeth Bernays transforms the structure and function of grasshopper jaws into poetry, metamorphoses hours of insect observations into applecart-upsetting scientific directions, and shares her life with passion and humility. It is an honor to view the marvel of insects through her creative magnifying lens.
—Marla Spivak, entomologist, 2010 MacArthur Fellow

Deftly, Bernays unravels the process of scientific inquiry, mindful of the capricious influences of gender, culture, and funding, but without losing sight of its joy. *Six Legs Walking* is the kind of book that will make you look at a blade of grass or a cactus pad with new eyes, searching for the "small wonders" that abound within its pages.
—Melissa L. Sevigny, author of *Mythical River* and *Under Desert Skies*

Six Legs Walking: Notes from an Entomological Life is a sparkling series of linked essays by a famous entomologist with a lifetime of close observation and experience behind her.

She speaks of "the image of insects, those primordial creatures that had been my passion for so long" and the effect the insects have had upon her life. Passion is the right word.

Her mother says, "I only want you to be normal, darling," meaning "like everybody else," but it is clear that this precocious child and brilliant woman will never be average and that her work with insects will elevate her life to a very special place and her writing to a very special level. A terrific book.
—**Richard Shelton, author of** *Going Back to Bisbee* **and** *Nobody Rich or Famous*

Read these essays and, like Bernays, you'll get obsessed with hawk moth larvae, leaf texture, the tiny feelers around the mouths of locusts, the sex lives of entomologists in London in the '70s, and the historical blight of the prickly pear in Australia, just to pick a few highlights. "Finding satisfying answers to unsolved problems has given me intense pleasure and is often why scientists become so absorbed in their work," Bernays tells us. The pleasure is shared, and it's yours for the taking.
—**Ander Monson, author of** *Letter to a Future Lover: Marginalia, Errata, Secrets, Inscriptions, and Other Ephemera Found in Libraries*

This book is an amazing way to fall into lives different from one's own—the lives of insects, of course, but also those of growing up in mid-twentieth-century Australia, of becoming a professional scientist when women just didn't do that kind of thing, of working in Mali, Nigeria, and India. The essays range from personal to scientific, often within the same page, with equally fascinating details about the lives of people and insects. Anyone with an interest in life, love, and death will enjoy *Six Legs Walking*.
—**Marlene Zuk, professor of ecology, evolution, and behavior, University of Minnesota**

Six Legs Walking should be in the library of all young ecologists; it is a wonderful example of one person's passion for science and the power of careful observation.

The writing is lush and sensual. You feel the heat of the desert, you see the colors of the flowers, you hear the tick-tick of her grasshoppers and smell the resinous creosote bushes after a desert rain. You feel her triumphs and her losses. Her final transformation is into a writer's writer. I invite you to enjoy her life's adventures as I have.
—**William E. Conner, professor of biology, Wake Forest University**

What an entomological life, with so much humanity! This book is literature and science, personal and professional, and filled with adventure and discovery. I recommend it heartily to all interested in biography, scientific journeys, and those beautiful bugs.
—Anurag Agrawal, James A. Perkins Professor of Environmental Studies, Cornell University, and author of *Monarchs and Milkweed*

Elizabeth Bernays and Reg Chapman imposed their mark on natural science and renewed the field of insect-plant evolution by introducing new partners in this duet—herbivore enemies, which we realize now are major contributors to evolution. In this touching book, Elizabeth comes back to the genesis of their ideas and tells us more about the backstage of this adventure, letting us peep into how these ideas emerged and permeated their life together. She also tells us more about her fascination with insects and passes to us her love of flowers and nature. A must read.
— Frédéric Marion-Poll, entomologist and neurobiologist

For those of us who already love insects, this book is a sheer delight; for those who haven't yet discovered the joys of watching our six-legged companions on the planet, Elizabeth Bernays is the ideal guide to the wonder of the unexpected observation; the deep satisfaction of exploration, experimentation, and discovery; and the ephemeral beauty of life.
—Martha R. Weiss, professor of biology, Georgetown University

From the evocation of a subtropical Brisbane childhood to lying in the mud at dawn in India observing minute caterpillars crawl the stems of sorghum plants to watching the ruthless killing by tiny predators and parasites as they shape the evolution of plants and the insects that eat them, Elizabeth Bernays has taken us deep into her remarkable life. Why would she and those of us who are similarly touched devote a life to such things? Read the book to find out.
—Stephen J. Simpson, academic director, Charles Perkins Centre, University of Sydney

Six Legs Walking is a captivating look at the study of insects through the eyes of a globe-trotting entomologist. Not only will you learn about the feeding habits of grasshoppers and moths, you will be drawn into deep reverie for the small wonders that populate our natural world while bearing witness to our interconnectedness.
—Gail Browne, University of Arizona Poetry Center

Six Legs Walking

Six Legs Walking

Notes from an Entomological Life

Elizabeth Bernays

Raised Voice Press

Clearwater, Florida

Published by Raised Voice Press
PO Box 14502
Clearwater, Florida 33766
www.raisedvoicepress.com

Front cover author photo courtesy of Elizabeth Bernays
Back cover author photo by Linda Hitchcock
Cover design by Melissa Williams Design
Interior design by Karen Pickell

Published in the United States of America

ISBN 978-1-949259-03-2 (Paperback)
ISBN 978-1-949259-04-9 (Ebook–Kindle)
ISBN 978-1-949259-05-6 (Ebook–EPUB)

Library of Congress Control Number: 2019939334

The following essays were previously published, sometimes in different forms:

"Color in the Tropics," previously titled "Aluminum Recollections in Nigeria," tied for third place in the 2009 Travel Writing Contest and appeared on the website of Transitions Abroad.

"Creosote Gold" appeared as "Creosote" in *Snowy Egret* 70, no. 1 and 2.

"Desert Time" appeared as "Time in the Desert" in *Eclectica*, Oct/Nov 2006.

"Flying with Lepidoptera" won the X.J. Kennedy Award and appeared in *Rosebud*, #39.

"Growing with Pollards" appeared as "Learning with Pollards" in the anthology *Shifting Balance Sheets* edited by Heather Tosteson, Kerry Langan, Charles D. Brockett, and Debra Gingerich (Decatur, GA: Wising Up Press, 2011).

"The Hawk Moth's Progeny" appeared in *Snowy Egret* 70, no. 2.

"Indian Medley" appeared as "Indian Pastiche" in *Driftwood: A Literary Journal of Voices from Afar* and also appeared as "Passage to Patancheru" in the anthology

Mambo Poa2 (Spain: Rezabal Taffet, Juan Roque, 2009), as a finalist in the Mikel Essery Travel Writing Contest.

"Jaws" appeared in *Snowy Egret* 68, no. 1.

"Cups and Nostalgia" appeared as "Plastic Cups in Arizona" in *Mary: A Journal of New Writing*, Summer 2011.

A portion of "Molting" titled "Arthropods Molt" appeared in *Fence*, Fall/Winter 2007–08, and was listed as Notable Non-required Reading in *The Best American Non-required Reading 2008*, edited by Dave Eggers (Boston: Houghton Mifflin, 2008).

"Prickly Pear Persuasion" appeared as "Prickly Pear Conversation" in *Antipodes: A Global Journal of Australia/New Zealand Literature* 28, no. 2 (2014).

"Rambling into Hungary" appeared in *Drafthorse*, Summer 2013.

"Sierra Interlude" appeared in *Summerset Review*, Summer 2006.

"Small Wonders" appeared as "Bugs" in *Umbrella Factory*, Issue 1: March 2010.

"A Taste for Novelty" appeared as "Variety and Black Grasshoppers" in the anthology *Blessed "Pests" of the Beloved West* edited by Yvette A. Schnoeker-Shorb and Terril L. Shorb (Prescott, AZ: Native West Press, 2004).

This book is dedicated to the memory of
Reginald F. Chapman

Contents

SIX LEGS WALKING

Small Wonders

I am nine years old. I run around in circles on the back lawn of our Queensland garden. Papa mows the grass and Mama plants the pansies. I like bugs. They jump up at me with all their colors; iridescent greens and blues, deep reds, yellows and oranges, striped and spotted, embroidered and painted. They grab my attention with big green wings, delicate shining membranes, black horns, clubbed feelers, jeweled torpedo bodies, feathery textures, lights blinking in the night. I think I am attracted to small things.

I like to watch the ladybird beetles, satisfying red roundness with black dots walking over and under leaves. They are like tiny mechanical toys as I let them walk onto my hand, and I sing to each of them as Mama taught me: *Ladybird, ladybird, fly away home; your house is on fire, your children are gone.* And more often than not, the little beetle opens its red domed wing covers, spreads out its thin black flying wings, and jumps into the air.

I see hairy caterpillars feed on the white cedar tree down on our second terrace. Sometimes they crawl down the trunk in long processions. The bold leader steps out, the next one has its head touching the leader, and the third touches the second. There may be thirty or more in line. When they reach the ground, the lead caterpillar takes them in what appears to be a clear direction, as if it knows exactly where it is going. Our white cedar still has leaves; there isn't another one for miles. I get a stick and direct the leader to turn left until eventually its head makes contact with the back of the caterpillar at the end of the procession. It stays there, still walking, and the whole tribe of them go in circles. Mama calls me in to lunch and when I come back out afterwards they are still doing it. The circle is about a yard across, and the light brown, hairy caterpillars keep moving all afternoon. They are still going when I am called to bed. Next morning there is no sign of them, though I search all around on the ground and up the trunks of nearby trees.

In summer, the oleander by the back door is covered with chrysalises hanging like silver ornaments on a Christmas tree. They came from striped caterpillars that feed on the leaves, and they will turn into butter-flies with whitish wings crisscrossed by lots of velvety black lines. Uncle Chisholm says they are called crow butterflies. The shiny chrysalises reflect the sunlight and, like tiny curved mirrors, sometimes create a funny picture of twisted leaves. Mama lets me cut a stem holding a chrysalis, and I put it in a jar in the kitchen so that we can see the butterfly when it emerges. I hope I will get a butterfly net soon. We watch it come out, Mama and I; the damp crumpled thing pushes out of the bulge at the top, then after a while turns upside down, holding on to the old shell, while it puffs and pants to spread out its wings until they are flat. It rests like that all morning before it begins to flutter. Mama says it is unkind to keep the butterfly in the jar, so we go out into the garden and open the lid. In a minute it is free and flies dreamily over the grass. It lands for a little while on a flower of the Quisqualis bush, and then disappears across the gully.

In the evenings I sit with Mama and Papa on the western verandah, where they drink a glass of rum and look out over the Brisbane River. Mama usually talks about the garden and what to do next. One December evening, she says, "It will soon be 1950." I suddenly realize that this means I will soon be ten, and that is a big thing. I often don't listen to them because my parents aren't friends and I hate it when they have a quarrel.

There are lights on in the house and, because we have no screens, moths fly in. They flutter round the lights, and sometimes the small ones get right inside the round lampshades on the ceiling, where they die. Mama has to take the globes off and empty them out. At night as I do homework, they land on my desk under the light and show off their greens and browns, yellows and grays. Some have long fringes on their wings; others have feelers like combs with long teeth. There is a common silvery one with white, furry legs and a long spine on each back leg.

The large moths land on the walls and just sit there most of the time, but the really big, brown bat moths fly around in the dark. Quite often they fly onto our beds and into our faces as we sleep. When this happens, I brush them away, and lots of tiny little scales called "goofoo feathers" come off. When we turn the light on, we see a gray patch of scales on the pillow or sheet. My sister Jennifer screams, "Goofoos, goofoos!" She hates these moths. Sometimes I get a chair and catch them on the wall for her before she goes to bed. I put a glass over a moth, then slip a piece of cardboard underneath and push it away from the wall. She wants me to kill them, but I don't like to squish such big things with wings that look like bits of lovely patterned carpet. I take them to the back door and let them go.

I do learn to kill the big grasshoppers that eat Papa's vegetables. They are huge—as long as my nine-year-old hand. They sit there in the foliage with their big bulging eyes divided into many tiny segments. They have no eyelids and they stare coldly. But if you creep up on them very slowly, they don't seem to notice and they sit quite still. When you get close to one, you can make a quick grab and catch it. You have to get it round the middle, because if you get it too near the front end, the long spiky back legs will kick and scratch your hand. If you catch one by a leg, it will escape, leaving just that leg in your hand. So you catch the grasshopper in one hand while you use pruners to cut it in half with your other hand. I don't like doing it, but Papa says it really helps him if I can do it, and, anyway, we have to save the plants. He says just do it quickly and don't look at it afterwards. There are good and bad insects and sometimes you have to kill them.

There are other killings I help with. I squash the greenfly covering the buds on Mama's roses between my finger and thumb, because she doesn't like getting the juice on her fingers. I help to pick inchworms off the lettuces and drop them into soapy water where they drown. I step on stinkbugs Papa shakes off the tomato and sweet potato plants. I spread flowers of sulfur on the azalea to kill masses of tiny mites with eight legs you can see only using a magnifying glass. Sometimes the sun's rays get collected in the magnifying glass so that they come out hot, and the mites fizz when they're hit.

The insects we all kill are mosquitoes. We often have mosquito nets over our beds and we burn mosquito coils to try to keep them away, but sometimes you just have to kill them. They like to hide under the bed and then, at night, they can get up under the net, so before we go to bed, we get the DDT sprayer and give the handle a few pushes under the bed. Mama says it's not good to use too much. Sometimes they are really bad under the dining room table, and we slap our legs to kill them as they feed. One night at dinner, my older brother Barton heaps up all the dead mosquitoes beside his place mat, which makes Papa laugh, and then we all start doing it to see who can get the biggest pile. Barton always wins things like this; he loves competitions.

I am eleven. At the new house, we race rhinoceros beetles. They fly onto the big western verandah with a heavy thud in midsummer, often falling like stones onto their backs and having trouble getting back up the right way. They are shiny and black with one or two horns sticking forward and curving upward. I hold a big beetle with my left hand and wind cotton thread round the horns a couple of times with my right, leaving inches of free thread on either side. Their spiny legs work overtime to get away, and they make a peculiar buzzing sound that makes my sister put her hand over her mouth and pretend to be sick. The matchboxes are ready with holes in them so that the cotton harnesses can be pushed through and tied in knots to turn each box into a cart, then Whoa, beetles, see how fast you can run across the verandah! They never take off and fly—I think they need to jump from high up to do that—but they walk all over, often not reaching the finish line I set for them. There is something about having a competition that really appeals to me, though I don't want to take part in any competition myself in case I am no good.

The rhinoceros beetles fly around the same time that the Christmas beetles come. Dozens of these shiny, silvery bugs come onto the verandah. Papa says that in the grub stage, they live inside the trunks and branches of trees and that probably this is why some of the gum trees have dead branches.

My school friend Leigh gets a load of silkworm eggs from a neighbor and gives me half. I am lucky because we have our own mulberry tree at home. I pick some leaves and put them in a shoebox with the eggs. I miss them hatching and at first only see holes in the wilted leaves. After that I "do my worms" every day. I take each caterpillar out and put it in a dessert dish while I empty the box and put in some fresh leaves. The caterpillars are whitish and smooth, and every now and then they molt, each time growing bigger.

One day I take my worms to school to show them around. Most of the girls say "Yuck" or "Pooeee." Miss Ramsay sees me looking in the box during class and makes me put the box in the corridor outside. Then I can't listen any more to what she is saying because I am worrying. The head teacher comes past the door, but she doesn't stop at the shoebox with holes in it. I feel relieved. At recess, I rush out to check them and then I take them to the cloakroom.

After about two weeks, they are over an inch long and they stop feeding. First they poo a lot and get a bit smaller, then they start spinning soft, cream-colored silk and cover themselves completely, making a little oval cocoon. The silk has to be wound off if I want to use it for something.

I wait a couple of days, then drop them into boiling water to kill the caterpillar inside. I get an old cotton reel and put a thin round pencil through the hole. It has to be loose. Then I hammer a nail into each side of the reel so that I can more easily turn the reel round and round. I have to find the end of the silk and start winding it onto the reel, but once that is done, I turn the reel, winding silk while the cocoon bobs about below, and, in the end, I get a reel about quarter full from just one cocoon. The idea I have is to sew something, but that never happens. The fun really is getting the silk from the cocoon.

I spend hours admiring butterflies landing on flowers, watching them open and close their wings. I see their long tongues uncurl when they land on the heads of lantana flowers, but they only stick them down into the yellow flowers in the middle, not the red ones round the outside. I guess the yellow ones have more nectar. Such knowing insects. How do they find this out?

Every Saturday morning, Mama and I enjoy the garden together, looking for butterflies as we ramble up and down the terraces, past the rose beds, along the path of the azalea walk, underneath weeping willows and jacaranda trees, through her patch of Australian natives, in and out of Papa's banana and citrus groves, down beyond the mangos to the chicken yard. I discover many types of caterpillars and learn to identify which ones give rise to which butterflies. I love every black and sky-blue Melissa flying high, green triangles closer to the ground, the gray-blue hairstreaks puddling in patches of mud, and the great birdwing butterfly with three-inch-long emerald-green and black wings and a golden-yellow body.

I don't think about anything else very much and I don't care if I am no good at schoolwork. I don't care if Jennifer says I am creepy or if Barton calls me a drongo or a dill. Mama and Papa know I love flowers and insects, so it's all right. I think Mama is happy that I am interested in something, because of my failure in school. I love to be with her and want to be good at something, especially because she and Papa are always sad or angry.

I am thirteen. For the school holidays, I go to stay with my friend Leigh at her grandmother's place on Moreton Bay at the mouth of the Brisbane River. There are oyster beds, a pier, and beaches. The land is very sandy, with tea trees growing in some swampy areas. Leigh's forty-year-old cousin Patricia comes to stay. She is tall with very thick, wild, short gray hair. She wears baggy khaki trousers held in at the ankles with rubber bands and loose cotton shirts beneath a sleeveless jacket with lots of small pockets. She can be rather fierce with us if she thinks we are silly, but she has a big, wide smile and speaks with a lisp, and she likes explaining things to us.

Patricia is an entomologist and professor, and she comes to collect tree hole mosquitoes. Tea trees branch a lot, and in the forks there is often a depression called a tree hole where water collects. She explains that some mosquitoes specialize in tree holes, laying their eggs in the water there, and it isn't known yet what all the species are or how the larvae, or wrigglers, grow and whether other insects or tadpoles get in there and eat them.

Leigh and I walk around with her and help. She has a special suction tube to capture the mosquito larvae, which she places in a preservative solution. She also has a little cup to take out a larger quantity of the water, which she puts in a wide, shallow white bowl so that we can examine the little beetles or worms or other larvae she collects. She keeps some of them and throws out the rest. She also keeps a sample of just the plain water to do some kind of chemistry later.

Patricia tells us all about mosquitoes. Some like salty water, some like it acid, others prefer neutral. They lay their eggs in groups called rafts. The larvae live by gathering up little particles from the water with whirling mouth brushes. At intervals, they need to surface to put a little breathing tube into the air to get more oxygen, and if you put oil on the water, it messes up the top of the tube and they can't get to the air. This way you can kill lots of them, which can be good, especially if they're the kind that carry diseases.

I think a lot about Patricia. I have never met anyone like her or even seen anyone wearing her kind of clothes. Mama always wears neat skirts and stockings with high-heeled shoes, and none of her friends wear pants or have such wild hair and no makeup. Patricia's baggy khakis are kind of funny, though I love the fact that she doesn't follow fashions or talk about boring things like weather and shopping. I love that she bothers to explain so much to us kids. I figure her doing all this stuff with mosquitoes is related to her not getting married, and I wonder if she cares. I can't imagine doing the things she does for a job. She teaches at the university, and we don't know anyone else there, and I know that for me and my sister Jennifer, the big thing we will do is get married and have children.

Patricia is like a schoolteacher. She doesn't like it if we don't pay attention when she wants to explain things. But she is also really nice and lots of fun, and Leigh and I love going on expeditions with her into the bush. After the holidays, she invites us to go on a weekend trip with the Queensland Naturalists' Club. She is president and has organized a visit to the Bunya Mountains northwest of Brisbane. We stay at a forestry station, in cabins outfitted with barbecues, carbide lamps, and double bunks. I learn that members of the club are called "nats," and that they are all kinds of people, most of whom have nothing to do with the university. Each of them likes a particular something in nature—rocks or birds, flowers or insects. Patricia leads walks in the forest, during which we look all about for birds and spiders and different kinds of mosses. Participants point out things they notice—the trapdoor of a spider, millipedes under logs, the fallen fruit of the bunya pines, along with the toe holes made long ago in the trees' trunks by Aboriginal peoples who climbed to harvest them.

Two entomology graduate students are on this trip, Ian from Brisbane and Saran Singh from India. They are collecting all kinds of small insects they find in the leaf litter, under logs, beneath bark, as well as in the vegetation they sweep with their nets. They look at every tiny specimen. To pick them out of the net, they have something called a pooter—a glass vial with two smaller pieces of glass tubing pushed through holes in a cork top. One of these has a short rubber tube attached, with another bit of glass tubing at the end. The other has a long rubber tube attached. You pull an insect into the large vial by sucking the end of the long rubber tube while placing the glass tube at the end of the shorter rubber tube beside the bug. Many insects can be collected in a pooter before it needs to be emptied into a killing bottle that has cyanide in it.

At the end of the day, when most of the nats are exchanging stories or examining bits of rock or snail shells, Ian and Saran are busy with their catches. They empty the killing bottles out onto big pieces of blotting paper and begin the setting and pinning. There are long, sharp pins for bigger insects, which must be pushed through the chest or thorax at just the right place. If the wings are to be set out, they are held with strips of tissue paper, which in turn are held down with short pins called lills. Very small insects are impaled by tiny pins called micros, which are then stuck into strips of pith on big pins; this is called double mounting. The minutest parasitic insects and flies are too small even for micros. For them, a tiny speck of special glue is placed on the extreme tip of a small triangle of card, the insect is touched to the glue, then the card itself is pushed onto a big pin.

We help Ian and Saran with their work. At the end of the trip, they give us starter materials so that we can begin our own collections at home. How very exciting it is to have a project! I quickly need more supplies. Patricia lets me have a box for storing specimens, Mama buys me a butterfly net, and Barton helps me make boards for setting out wings of butterflies and moths. Now my homework at night by my lamp is mixed with moth collecting and setting, and my weekends in the garden and bush are busy with searching for new specimens.

At school over the next few years, Miss Gray encourages and helps me. I learn the orders and families of insects on my own, and I spend hours making tiny labels with a fine black pen. I send specimens to the National Museum of Australia in Canberra. Many of them are new to science and they are kept for the national collections, but if there are duplicates, I get those back with the new names. I am in love with zoology, and especially insects; I watch behaviors and learn about evolution. I enter the Queensland School's science competition and win a prize, my first entomological success. Miss Gray is so inspiring. She makes so much more sense to me talking about Darwin than the Presbyterian and Methodist ministers do during our Bible classes. I think because of her I do well in the public exams at the end of my final school year.

I am eighteen. I get a scholarship to go to university. No one in our family has ever been to a university, and Mama has the vicarious pleasure of my excitement there. She is really clever but never had a chance to go to college herself. She married my father only to please her parents, who would not accept the man she really loved. I will become an entomologist, though, and Mama will be so proud. Now I feel sure that during all those years of bug love, Mama took her joy from my joy, and the butterflies soothed her twice over. Her mournful happiness is mine now as I seek to take in nature as deeply as I can in this short life. I think, even back when I was nine, the attraction of flowers and insects was comfort against my sadness, which was somehow related to Mama's sadness.

Flying with Lepidoptera

The house we lived in when I was nine was built on wooden stumps that were at least ten feet high on the northwest corner of the house but less than two feet high on the southeast. A wide verandah looked west out through gum trees and over the Brisbane River to the open fields of Fig Tree Pocket; stairs led down from the high verandah to grassy slopes with palm trees and, eventually, a driveway. My parents sat on the top two steps on summer evenings smoking Craven "A" cigarettes and drinking their noggins of Beenleigh rum. Smoke from a mosquito coil curled round the railings, its glow becoming fiery as darkness fell from behind the house. Across the river, dingoes howled. I sat on the step below Mama, whose left hand played with my curls as she talked and talked about the past, her family, and Queensland history, while Papa sat silently gazing into the distance, offering a "Mmm" or "Yes" as needed.

Occasionally, a big, dark velvet moth flew by us and into the house through the French doors. During teenage nights, I collected the multiplicity of moths that came to the circle of light made by the lamp on my desk. Silently, during homework hours, I put moths into a killing bottle; later, I tipped them out and set them each with a pin through the thorax into a trench in the setting block, wings spread and held out with strips of tracing paper and pins onto the shoulders of the block. I drew each antenna to the forward position, placed each leg outward. Some of the tiniest moths were beauties, with silvery scales and fringes along the trailing edges of the wings or delicate patterns like watermarks on paper. My favorites, the plume moths, had deeply bisected wings covered in delicate cream scales making each wing resemble a minute, ghostly palm leaf. Through my hand lens, I gazed at each handsome scrap of nature's intricacy.

At the back of the house, a couple of steps led down into a colorful garden of terraced lawns, wandering paths, and flowerbeds, all of it developed and maintained by my parents from a wilderness of weeds. Beyond the garden, the wild bush crept east up the hill, and through it led the "goat track," which Papa and Barton and I trekked up in the mornings, reaching Dewar Terrace at the top and continuing down the other side to the train station. Another path from the back gate, the "kangaroo track," led north into a valley and up the other side to Hilda Street. This was the way the milkman came to deliver, into a waist-high box, our daily half-gallon of milk—first in a bucket and later in milk bottles. This was also the way I went to talk to our closest neighbor, Mrs. Johnson. When my little brother Adrian was old enough to go to school, he, too, went by the kangaroo track, secretly leaving his shoes behind a tree so that he could be one of the bold barefoot boys.

We had an oleander tree by the back steps, where I watched conspicuous caterpillars, striped black and white around the body with long black feelers. I loved how, when they finished their feeding lives and changed into pupae, they became silvery chrysalises hanging under leaves like botanical jewelry. As the slanted morning sun caught the curve of one, another sun shone from its surface, lighting my childish days and shining onto a future of butterflies that I would catch or watch, butterflies that hovered on the horizon of my consciousness, lifting my thoughts into new layers, connecting threads of ideas into a web of faith in life itself.

On certain warm, sunny days, a black-and-white butterfly slowly emerged, breaking its chrysalis, bulging out damp and limp, then blowing itself up and hanging there drying before flying off along waves of summer searching. My much older brother Barton, who knew everything, said, "Don't touch them, they're poisonous." Jennifer said, "Little sister, those caterpillars are just disgusting." Papa laughed at me for watching pests eat one of his cultivated bushes. I looked at them without touching as I sat on the steps, but my interest in the silvery chrysalises and the clever emergence of those delicately tough-winged creatures I couldn't catch was complete. Mama agreed that the transformation of a stripy caterpillar into a silvery chrysalis into a delicate black-and-white crow butterfly was infinitely strange. One day she warily cut off a leaf with a chrysalis and put it in a jar by the kitchen window so that we could watch the butterfly emerge before giving the fluttering captive freedom in our garden. Later, I would learn it was a species of Danaidae, the family that includes the monarch butterfly.

As an entomologist, I studied a pest grasshopper that defoliated the cassava crop in Nigeria. It was a multicolored thing called *Zonocerus variegatus*, and because of its bright red and yellow coloring on a black background, I assumed it was aposematic—that is, warningly colored and containing chemicals toxic to predators. I had no solid evidence, however, as no one had studied its chemistry or knew if predators avoided them. It was singularly susceptible to fungal disease, though; as death approached, the infected grasshopper climbed to the top of a bush and expired in the early evening, usually about seven o'clock—a perfect place and time from which to disseminate fungal spores to other victims.

One morning before the dew had dried, I was out counting live and dead grasshoppers in a cassava plot for our population studies when I saw a queen butterfly, *Danaus chrysippus*, alight on a corpse, tap it with tiny feet, and touch it with lowered antennae. For a minute he (for it was a male) hesitated and opened his wings flat towards the morning sun, then began probing with his unfurled proboscis all over the dead grasshopper while intermittently lowering his antennae. Eventually he reached the anus, from which exuded a viscous dark liquid, and there he drank avidly from the liquefied gut contents dripping out onto the leaf. Immediately, I realized that the eager butterfly had given me an important clue: Like the monarch and other danaids, this butterfly species takes up special toxins called pyrrolizidine alkaloids from withered plants to make their all-important courtship chemical. But getting the chemicals from plants is a difficult task; the butterflies must secrete saliva onto the plant surface in order to dissolve the few crystals of toxin that occur there, and then they must suck the liquid back in, repeating this process many times to obtain the right amount of the chemicals.

Perhaps, then, here was a ready source of the important toxins derived from plants eaten by the grasshopper in the days before its death. This discovery led to experiments in which I fed known toxin-containing plants to grasshoppers, found a chemist who analyzed the insects later, and discovered that, indeed, *Zonocerus* sequesters the pyrrolizidine alkaloids and thereby gains protection from various predators. Furthermore, they use a breakdown product of the chemicals as an odor to attract more of the same kinds of grasshoppers, resulting in groupings of males and females at special mating and egg-laying sites.

Mama and I enjoyed our Brisbane garden (she always called it "the estate") together every Saturday morning, looking for butterflies as we did our rambles up and down the terraces. She was pleased for me to have an interest. Papa didn't mind that I was always at the bottom of the class, and, anyway, he never talked. He just liked having his "flower fairy" in the garden, helping with the cultivating and planting, a little girl he loved to smile at and tease. But Barton was scathing. "Oh, poor old Punka, you're such a dill," he would say. "Elizabeth is real drongo." "Fair dinkum dimwit." "A shingle short, I reckon."

I attended a Presbyterian-Methodist school for girls, where I received the lowest grades in all the tests. I also found the occasional questionnaires baffling: What did my father do? What denomination was I? Where was I born? I was never sure of the answers, so I just kept quiet. While watching butterflies, however, I put school out of mind. Silver chrysalises and butterflies were my secrets. The natural world, the small things out there in the garden with Mama, the little blue butterflies that rubbed their finely penciled back wings together, prevented me from being sucked under by all the people who agreed that I was slow.

The blues, or Lycaenidae, were my favorites in all the order of Lepidoptera for their daintiness. And they still are. Some years ago, I studied them in Colorado, near Gunnison at the Rocky Mountain Biological Laboratory, with Dave, my star PhD student at the time. We examined two types of blue butterflies: one used a single lupine species as its host plant, the other used many species of lupines. Why the difference? We followed egg-laying females in the field, monitoring their choices before and after landing and examining the survival of their larvae. Dave did most of the work, returning for several summers; in the year I went, I experienced the joy of simply looking. My eyes took in all of the enchanted alpine world, with its flowers and butterflies, but keenly watched the two blues. The specialist fed on leaves, a tissue available throughout the summer; the generalist preferred flowers, which are short lived and unpredictable, since flowering times vary with the weather. Clearly, it would be madness for the generalists to have only one host-plant species. The story became more complex for Dave as his research continued, but at the beginning, I enjoyed trailing the little blue butterflies as they, dreamlike, rose up and floated down, savoring floral exuberance; drinking nectar from purple thistles and yellow daisies, from dark pink fireweed and pale pink phlox; laying eggs on blue-flowering lupines. Somehow, the watching of butterflies threaded through my life and gave me joy independent of the fun of discovering some detail of their lives that could be translated into answers to larger questions.

It must have been when I was thirteen that Mama read in the paper about the new vocational guidance center in Brisbane. The field of psychology had grown in the early 1950s, resulting in new jobs and new services. The center carried out intelligence tests and advised about career options. Mama, worried about my slow progress at school, was determined to try this crazy new thing. We dressed up; I wore my navy-blue school uniform and Mama wore her gray silk dress, new hat and gloves, and black high heels. She always liked to be early, especially in summer, so that she could take time to wipe the steam off her glasses and gather her wits. She perspired both in the heat and due to anxiety, and this day she was on edge. She wanted me to be normal. And I wanted to make her happy.

We waited together in silence. Then a tall young man with a mustache took me to the tiny test room. Through a window, I saw an older man sitting in the next room. The young man was gentle: "Okay, sheila?

Orright. Now here's what you do. There's thirty tests, and you start with number one and just go through them. You need to work fast, Okay?"

I nodded.

"Every time the bell rings, you stop the one you're doing and pass the sheet through the window to that bloke and start the next. Okay?"

I nodded.

"Okay, sheila, when the bell goes, you just start."

First there were quizzes with words, then math, then shapes. Every time the bell rang, my heart pounded and my mind went blank for a while. I could feel the man in the window watching as my mental capacity faltered and my bare sweating legs stuck to the metal chair.

It was a long wait with Mama afterwards before we went into an office for the results. There were beads of sweat on Mama's upper lip.

"Unusual," the man said. "Hard to interpret."

There was a pause, and Mama started to get that forced expression she put on for photographs. Eventually, he gave his speech. I can remember a few words and phrases: shorthand typist; don't bother with exams; no, not even nursing, I would say, subnormal actually, but with discrepancies.

My passion for butterflies increased. I ran for Melissas and birdwings with my butterfly net and crept among the bushes to find their caterpillars. I started a collection, and my Uncle Chisholm gave me *The Insects of Australia and New Zealand* by R. J. Tillyard.

Mama would say, "I only want you to be normal, darling," so I tried to make it up to her by working at the things my sister shone at, such as tennis and sewing and going to dances. But somehow I never understood the rules of combat or the things that mattered by the book of fashion. Still, though, I felt she loved me more than she loved the other three. After school, we worked side by side cultivating the delphiniums, trimming the lawn, pulling weeds, a closeness with few words. It was at these times that she gave me her few words of instruction for life: "We belong to the Church of England, remember that." Or, "Don't listen to those bigoted Baptists down the road, because all that matters really is the Sermon on the Mount."

I sometimes watched her work from the kitchen window as I made her cups of afternoon tea. At intervals she smiled at the yellow butterflies, and then my afternoon became the lifting flight of butterflies as they fluttered round the blue and orange flowers. It is the yellow butterflies now in my Arizona garden that remind me of those singular days, when the sight of kneeling Mama with her unknown sadness and sudden smiles became fixed in my memory. It was only later that I realized her losses—the love of her life rejected by her parents, marriage to a man who could give her so little beside debt, and the great loneliness that ensued.

Jennifer left home to become a nurse when I was just ten. Her life was one of evening gowns and makeup, of being in love and playing waltzes on the piano, an indoor world I knew little about. I put up with all the strange things Barton made me do and took the many presents he bribed me with, but he was also in another world, of motorbikes, cars, sailing, and woodwork. Adrian was just a baby. Mama and I, though, we had our shared life. For all her dislike of creepy-crawlies, she was with me in my love of flowers and butterflies.

On rainy days, she read to me from Arthur Mee's *Children's Encyclopedia* or from *Grimms' Fairy Tales*. But best of all, we looked at *Homes and Gardens* magazines together and imagined all the projects we could do in the garden: make paths of stepping-stones, build a pond or a rock garden, grow creepers over the wall. In one issue of the English magazine, there was a painting of a butterfly almost identical to the American painted lady butterfly on a postcard from traveling Aunt Muriel, and I was surprised to realize that the same kind of butterfly could occur all around the world. In fact, *Vanessa cardui*, as entomologists call it, is the most abundant butterfly known.

Even now, butterflies remind me of how I changed from a backward child into a scientist. I ran headlong into success, though the transformation took years to achieve and to believe. I was lucky to have schoolteachers who convinced me I was normal, and, in my resultant adoration for them, I found my feet. Prohibiting failure, they insisted I combine my love of butterflies with conventional achievement. Their victory, though, was fully accomplished by the thrill I had surprising Mama, by the joy it gave us both to find that I was not a dunce, that the vocational guidance men had been so wrong, that so much had been wrong.

Academic achievement as a teenager allowed college. Then Mama's vicarious pleasure at the novelty of my university experience was yet another reason to enjoy my continuing progress, until finally I could say, Mama, I am a technician, I am a graduate student, I am a scientist, I am an entomologist, I am a professor.

The thread of painted lady butterflies wound though my life. In England, I was delighted to see them in the heathland of Sussex as they drank nectar from dandelions and thistles; I walked from village to village with a book of poems and a light tread. I found them at the Oxford University Museum of Natural History in E. B. Ford's collection of butterflies with bird bites out of their wings. In Switzerland one early summer, I saw them migrating north; perhaps they would go as far as Norway. In Kenya, I watched a male patrolling for females on the Serengeti Plain.

Here, in my Arizona garden, I see *Vanessa*, the painted ladies, regularly. They perch on the flagstone, visit the yellow *Viguiera* flowers, and stop by the small pool where birds and coyotes drink. As a professor at the University of Arizona, I advised a student, Dana, who thought avoidance of natural enemies might be a reason why herbivorous insects use specialist host plants; his model system was *Vanessa cardui* in the American West. Every spring migrants come north from Mexico, and, here in Arizona, they lay their eggs on just a few plant species—thistle, mallow, and lupine. In good years, after sufficient winter rain, these plants are abundant and can support big populations of the caterpillars.

In one year of his study, they were especially numerous and seemed to just keep coming—there were days when the air was filled with the purposeful floating flights of the ubiquitous and beautiful butterfly, one of evolution's shimmering phenomena, a species present on the earth in billions, each individual offering a potential epiphany, a passing moment of exploding emotion, a reason to stay alive.

Dana studied *Vanessa* butterflies in Arizona, where a new set of larvae grew to butterflies and flew on ever northwards to Colorado, where he watched females again lay eggs. Another generation of the fresh, north-flying creatures reached Washington State, where he watched them yet again. In the fall, they would go south, a ragged and less conspicuous return.

Why did their host-plant choices change in successive generations northwards? In the end, it seemed that different predators applied different selection pressures in each habitat. For example, here in the desert, where ants are such a hazard, larvae are found mostly on thistle, where it is possible for the caterpillars to build up a silk nest in a leaf, draw the edges together, and face the ants with rows of impenetrable spines. In more northern places, lupines are better hosts because the alkaloids the caterpillars get from eating lupine plants protect them from certain birds.

The story becomes involved, and we still don't know it all. Success cannot be denied, however. *Vanessa* is beautiful, and highly adapted to the versatile life of a worldwide migrant—it is opportunistic, high flying, flexible; it escapes vicissitudes by moving on, not held back by bad weather; it escapes enemies, it explores, it finds what it needs. And that undulating flight, so deceptively absentminded, is as consistent and determined as waves on the ocean, as sure to arrive as the sun in the morning. And so, I rejoice in *Vanessa* and all those scale-winged beauties that represent life lived, life among the flowers, my life fulfilled for Mama.

Molting

In order to increase in size, each spider, crab, scorpion, and grasshopper must shed its unstretchable exoskeleton—a process refined over a period of six hundred million years. To watch a molt is to see something primitive and prehistoric. It is difficult and intricate, yet it can be observed on any summer's day.

A grasshopper walks slowly towards a tree trunk. He seems a little less agile than normal. He has not eaten for a day; his digestive tract is empty. He climbs up, then settles the claws at the tips of his feet into irregularities in the bark and rests. He is quite still except for the smallest breathing movements. After an hour or so, he appears to pant: his segmented abdomen extends and retracts like a concertina. He is drawing air in and out of holes along his sides that lead to branched breathing tubes and respiratory sacs. Similar to an athlete, he is preparing for his great act, though his preparation is lengthier. The panting becomes faster, more resolute. His helmet-like head juts in and out on a flexible neck. His air sacs are cramped by the compression of his tissues, waiting. After a couple of hours, he becomes completely still.

The panting phase over, he swallows air. We can't see it, but his abdomen extends; his head pushes out, forward, and back, extending his neck membrane. The rudimentary little wing pads on his back swell, forcing them apart. With an X-ray machine, we would see his whole digestive tract balloon out to the sides of his body, squeezing tissues flat and greatly increasing the pressure of the blood that presses against every part of his external skeleton. We might barely discern a shadow of change in the shape of his thorax, when suddenly—along a line of weakness up the middle of his back—the old cuticle splits, and soft, new-made folded matter that will later become hard cuticle pushes out due to the pressure from within. There is a pause while he swallows more air. The rupture, with the emerging thorax extended forward along the top of his head and backward to the end of his abdomen, creates a bow shape as the upper part of his body expands while the lower part is still enclosed in old cuticle. The plump, crumpled wings are pushed out and his face is pulled down. With strong neck muscles attached to the new exoskeleton, he lifts his head, pulling it half out of its case. He wears a gas mask of thin old cuticle. He stops only briefly, as he must continue before the lubricating fluid between the two cuticles dries.

A new movement begins. With tiny, deft rotations, he begins to fold the old cuticle onto two spiny protuberances at the end of his abdomen, steadily drawing the old cuticle back and away from his down-turned head, allowing more expansion at the front end of his body. His still-encased antennae are pulled backward, but he can now draw out his front legs to grip the empty shell, which remains firmly attached to the tree bark. He struggles to pull his head up; he puts one front leg on an antenna ready to be withdrawn from its case, then he deals with the other antenna, and lo! His head is up and free. From then on, the remainder of his task is simpler. He pulls, and the backward folding continues, bunching up old cuticle at his rear end. All six legs are freed. He pulls out the cuticular lining of his branching respiratory structure—a bush of limp, silvery branching tubes. Standing on his cast skeleton, he continues his expansion. More air is swallowed. Soft and helpless, he pumps blood into his wings, which expand greatly, becoming filmy. The hardening has already begun.

During the final curing process, he lets out the air from his gut, allowing his blood pressure to return to its normal low level. And as he does, he pants a little, gently extending his body forward and then back to fill the delicate air sacs, for now there is space to become fully aerated, loaded with oxygen, and lightweight enough for running, jumping, flying. After an hour or so, his new cuticle acquires its proper color as it hardens. He stands tall, his new, firm wings fold, his antennae point forward. Then he walks away as if nothing at all has changed. In another hour, he resumes his life of feeding, his mandibles, with their newly sharp teeth, now hard as chisels. He bites through a blade of tough grass.

Two years had passed since I left my Australian home. I was in England and ready to return to the magic of insects—those primordial creatures that had been my passion for so long. Insects with all their diversity and colors and fabulous molting behavior called to me.

During my days at the University of Queensland, with all the excitement there, my life had been changed forever by all that was not entomology. The liberal camaraderie at every level was an eye-opener after my strict high school, while the freedom and endless ways of knowing were staggering. Above all, there was the Bushwalking Club. On bushwalks, I was nourished by wilderness and the silent enjoyment of forests, mountains, and far horizons, but I also learned the meaning of shared ideals that were not rooted in religion. I learned about discipline under duress and made lifelong friends who would have a lasting influence on me. I became a regular for the weekend bushwalks. There were never paths. We made our way with maps and scorned anything that was easy, such as solid fuel or fancy tents. After long, hard climbs, I would lie amongst the fallen leaves looking up at the *Eucalyptus* trees with scudding clouds beyond, noticing that every leaf had insect damage, though it was not the time or place to share my observations.

Dozing off during dreary lectures in those undergraduate days, I dreamed of childhood and the joys of collecting jewel beetles, butterflies, and moths, of lying out in the family garden gazing at the feathery leaves of the jacaranda tree, and of rowing my pram dinghy across the Brisbane River to Lone Pine Koala Sanctuary. Though I would graduate with honors in zoology and entomology, different doors had opened to me, and the astounding vista of other possible worlds had impinged upon my conscience. During a long bushwalk across Tasmania with my college friend Lucy, we hatched plans for something really exciting: travel by ship to the other side of the world.

When the time came, I was twenty-two. I had a one-way tourist-class ticket on the beautiful P&O liner *Canberra*. At Circular Quay in the middle of Sydney, the huge ship seemed as incongruous as an iceberg in a tropical forest, with its giant lianas of ropes tethered below. When at last I climbed the gangplank, I threw my colored-paper streamers from the high deck into the thickly milling crowd below, and those little people we knew down there caught them. Then lianas were undone, and the bobbing tugboat pulled us away. All the streamers tore apart and fluttered in the breeze as we sailed through the rocky heads at the entrance of Sydney Harbour, with the bougainvillea-red sunset behind us.

And so began a year of travel, mostly in Europe, followed by another year of bohemian life in London before I finally settled down to teach biology in the tough Cockney East End. It was a delightful new challenge to learn how to engage those bright young people in the workings of bodies, including their own. How I laughed, too, at all the children's self-deprecating humor as I learned their rhyming slang. "Love yer titfer miss" (titfer meaning "tit for tat," replacing the rhyming word "hat"). "That's me china" (china plate, standing in for "mate"). "He got a lotta cobblers" (cobblers awls: balls).

In spite of all the fun, I missed learning and I missed insects. I searched for opportunities and chose to spend two years doing a master of science taught by Dr. Reg Chapman at Birkbeck College, part of the University of London. Our classes were in the evenings, three nights a week, as the program was designed for people in full-time jobs. It was a rigorous training—a five o'clock lecture followed by four hours of lab work. And usually after class, we all went out to the pub with our ever-friendly teacher.

"The only really social animals are ants and bees. Humans are certainly not—except perhaps the Chinese." This was Reg in provocative mood. It was late, and the class, all eight of us, were swigging beers at The Lancaster Arms in Bloomsbury when the outrage began.

"What's *your* definition of social, then?"

"You mean you put *insects* in a higher position?"

"Are you in favor of *Mao*—the little red book?"

"Ants are just automatons, *that's* not social!"

"And bees are just run by *pheromones!*"

"You have to distinguish between social conditions with free will and *automatic* social behavior!"

I could see Reg grinning to himself as the arguments raged. He loved to get people going this way, and, as usual, everyone had fallen for it. I had, too, for a few minutes. More beer was needed. I bought a round. The gathering didn't break up until closing time at 11:00 p.m. As the barman called, "Time, gentlemen, please," John got up to buy the last drinks, ensuring we each had bought a round and had drunk several pints. We went our separate ways, each of us in a different state of mind—amused, angry, dissatisfied, but, at any rate, thoughtful. And that, it seemed, was Reg's aim.

In the class of 1965–66, four of us were young teachers; besides myself, there was Little John, Mark, and Dave. Edward C. was a forty-year-old chemist from industry, Edward T. a mathematician close to retirement, and Big John a young technician. Dinah, with the bright red lipstick, had some job in Reading we were never told about and was of unknown age, though certainly under twenty-five. Reg was in his thirties and a Cockney who had made a name for himself in entomology.

I generally arrived straight from Raine's Foundation Grammar School and took a hurried early dinner in the basement cafeteria, where warmed metal containers held roast potatoes, peas, carrots, slices of beef, pieces of fish fried in batter, and stew that seemed to have been there all day or longer. Sometimes Reg was there, too, and we ate together, but it wasn't a place to linger, with its dim lights and overpowering smell of old fat. A better place was the library, where I often spent time before class began.

On one side of our little classroom/lab was the office of another lecturer, Old Tom. He usually stayed in his office until after our class began, and Dinah often spent time with him before class. I could hear her chirpy, high-pitched voice and shrill laugh behind the locked door as I passed, and I wondered what they were up to. On the other side was Reg's office. It was filled with equipment, though he had use of the classroom for his research during the daytime when all of us students were at our jobs.

Before class, some of us might visit Reg in his office to ask questions or have discussions with him, but more often we gathered in the lab next door and gossiped. Did the department head fancy his cute little technician? Had anyone seen that botany lecturer expose himself? Did everyone know he was kicked out of the Gloucester Arms pub? Dinah talked about the virtues of springy mattresses. It was not explicit, but we all knew she was talking about sex, with her cherry-red lips as she fluttered her eyelids and tossed her head. Dinah fascinated the men, and she fascinated me, too, with her sexy manner and clothes and her very quick wit.

One evening in the summer of 1965, Dinah and I were alone before class, and she pranced up and down the lab in her new scarlet, pink, and gold minidress straight from trendy Carnaby Street; she was dressed, she said, for a Rolling Stones concert later that night. She surprised me by coming close and saying, "Reg is pretty attractive, don't you think? Well, sexy, in a funny sort of way?"

I hadn't really thought of Reg as anything more than the lecturer who taught our classes, but after that, I began to notice more about him. He usually wore old gray trousers, a white shirt with a college tie, and a gray sweater. After he started lecturing, he would take off the tie. Then, a bit later, he took off the sweater, revealing sleeves rolled up above his elbows. His worn belt was tight, just below a faint paunch. He got excited when he was teaching; he lit up when we asked questions. He loved his subject and he loved explaining everything. That passion, I think, was his greatest appeal to me. And I liked his plain face, with a large mole on one cheek, a slightly prognathous jaw, and unflattering glasses. He was not attractive in any conventional sense, but he was such a pleasure to watch as he shared his knowledge. For me, his classes were never long enough.

Birkbeck College began in 1823 as the London Mechanics' Institute and had flourished from the time of its origin, opening up the worlds of art and science to many ordinary Londoners. The "mechanics," as skilled artisans of the day were called, were an eager audience for physical sciences, politics, languages, and music, though the idea of education for all classes sent a shiver down certain establishment spines. Some even accused founder George Birkbeck of "scattering the seeds of evil." Later on, he was hailed, and loved, as a champion of popular education and was held in great esteem.

The Institute was opened to women by 1830, and in 1892, Millicent Fawcett, later a well-known suffragette, lectured there on the problems of poverty. As it became part of the University of London, Birkbeck College remained a place of higher learning for people unable to afford full-time study, and it attracted faculty with a concern for less-advantaged students, including women. Dame Helen Gwynne-Vaughan joined the Department of Botany as the college's first female professor in 1909, and today, Birkbeck still boasts relatively more women professors than other colleges. In this place, I joined the ranks of women inspired by academia. I was proud to be attending Birkbeck College of the University of London, pleased to have an interesting teacher in Reg, and rather excited to be back in a university setting after time out for a couple of years. It was also a stimulating contrast to teaching school kids who were entertaining but not so interested in learning.

I liked the busyness of the dull building at night and the feeling of so many committed people who, like me, wanted to learn something after a day's work. I liked being in classes with other people who brought diverse backgrounds and experiences with them, who were stimulating to talk to during the laboratory classes. I also liked the way the school retained some feel of its origins, a socialist facility for a largely untutored populace. The crystallography genius and Marxist J. D. Bernal was still there, as was Hans Eysenck, the brilliant but controversial psychologist.

One day, Reg and I had a pub lunch and afterwards walked to Bloomsbury Square, where we sat on a bench for a while. We talked about some entomological thing that I don't remember, but what I do remember is how much I respected and liked him. When we stood to go back to Birkbeck, he turned to my admiring face and our eyes met for more than the usual time. Then we kissed, and it was the beginning of something strange and new. All those stories of intense love had seemed like a chimaera until that day. Leaves of formless shrubs turned to luminous jade and the crystal chilly air filled with tiny specks of diamond light; the very stones seemed gilded though there was no sun. Nothing was ever the same. Reg and I were soon deeply in love.

In the last summer of the master's program, a summer of love among the grasshoppers, Reg employed me to do museum work with the collections and to help with experiments in the field. He wanted to know how selective English grasshoppers were in their natural feeding habits. We analyzed the proportion of plant species in a field, and we studied under the microscope the minutia of leaf anatomy as well as the droppings of grasshoppers we had collected. Leaf cuticle is preserved intact as it passes through the gut, so by examining the droppings, we could identify the plants eaten and then determine whether the grasshoppers favored certain plants relative to their abundance. In fact, they did, and this work became my first entomological research paper with Reg. He put my name first on the paper, insisting, "I always just put the names in alphabetical order."

I knew Reg appreciated my entomological knowledge and my enthusiasm for the subject. In writing his monumental text *The Insects: Structure and Function,* he used me as his student reader for each chapter, meaning that I had to point out anything that was not clear. I was conscientious and took it all in. I enjoyed our discussions about the writing and, in the process, became completely reinvigorated in my interest in insects. How complex, how well adapted, and how diverse insects were. What fun it was to be able to take part in the book project, and how encouraging that Reg thought I had something to offer.

He wanted to continue work with me, he wanted me to succeed, and he wanted me to study for the PhD with him (though others would have to be my examiners). As well as my love for him, I felt deep gratitude at being wanted for my intellectual companionship. Insects, which had been so fascinating to me as a child, had once again become the center of my life. The entomology I had first gone to university to study had once again become exciting. And so, I chose to stay with Reg. As I worked on my doctorate, trying to solve biomechanical problems and working out the musculature of hatching and molting insects, I had my close friend and lover in the next room, my entomological inspirations mixed with admiration, my achievements aimed at pleasing Reg even more than myself. And he was fun in so many ways, including his use of rhyming slang. If I did anything thoughtless, he would laugh and say, "Lizzie, use your loaf" (loaf of bread: head). If he wanted me to look at something under the microscope, he would say, "Take a butcher's" (butcher's hook: look).

My chosen work was concerned with hatching and that extraordinary process of molting in locusts. I took the tube train each week to Kensington and obtained small aluminum cans of sand from the Anti-Locust Research Centre, each of which contained locust egg pods. Caged females had dug deep into the sand and deposited sixty or so eggs, each one slowly retracting her abdomen as she laid them, forming a cylinder of eggs a couple of inches high. Then, as she emptied her load, she continued shortening her abdomen while at the same time depositing a channel of white froth above the eggs. The froth stiffened just a little, providing a future route through which hatching young could easily dig up to the surface. In time, I managed to get females to perform this feat in sand between sheets of glass so that I could watch and film the digging process.

To start with, though, I wanted to find out how a hatching locust emerged from the egg. I emptied the sand and separated the eggs onto damp filter paper in a small glass dish and noticed that, as the time for hatching drew close, the embryo became a tiny, squished locust made visible through the eggshell after the egg was washed in a little chlorine solution. A few black, hard spots—dark jaws and leg joints—revealed the little creature as being almost ready. I kept the dishes in my warming cupboard at home and set my alarm clock to ring each hour so that I could monitor the timing of each change more precisely. When an egg was close to ready, the enclosed nymph would begin moving if aroused by light or touch, and then I watched the quarter-inch-long football-shaped egg with a microscope.

The process was always the same. The tiny locust drew up its abdomen and expanded its thorax sideways, in a rhythm that lasted an hour or two until, finally, the shell split across just above its neck region. It was one thing to watch, but then other questions arose for me: What pressure was being developed? What muscles were being used? Why did the shell always break in the same place? Was it simply from pressure or was there some other mechanical aid?

These questions were answered in time. I built a miniature pressure gauge, and I studied the detailed structure of the little nymph by making hundreds of slices that could be viewed microscopically and drawn. From such drawings, I could reconstruct the whole insect, enlarged and in three dimensions. Getting out of the shell was mainly a matter of pressure developed by contraction of the abdomen and by special hatching muscles that changed the shape of the thorax. The neck region had two protuberances, or ampullae, that were filled with blood under pressure in a contraction phase, and these pressed outwards and eventually split the shell across the neck. I later found out that the ampullae and the hatching muscles disappeared in the days following hatching; their work done, they were never to be needed again. The efficient little baby locust could reuse the materials.

After hatching, the little insect must dig to the surface of the soil through the froth the female deposited after laying her eggs. To study the digging phase, I had to leave the egg pod and froth intact and study the process through glass. I managed to get hold of a Bolex cine camera; it was long before digital photography. These were also the days before one could film in low light intensity, so I had to devise a means of brightly illuminating the little creatures without cooking them. I focused a bright light through a water-filled round flask so that I could enhance the intensity of the light at a small point, and through the flask I ran a constant stream of cold water so that the heat could be absorbed, allowing me to get very bright, cool light. In this way, I made movies of the little insects digging, which I used for detailed analysis later.

The movements of the tiny creature in the froth involved the same muscles used in hatching, but because the nymph was no longer in a confined space, the effect of their contraction was different. Abdominal contraction pushed out the ampullae into the froth and sand, making a purchase, then further contraction pushed the head upward into new space. The wormlike progress was aided by the nymph's pointed head shape, kept that way by the constricting embryonic membrane tightly fitted over the soft, wrinkled head cuticle beneath. The work seemed difficult and the rhythm endless, making me feel tired myself. In nature, the process takes perhaps ten minutes, but in one experiment, I made a path for the baby locust that was circular, and the poor creature went round and round in circles for hours. Apparently, the hard-working insect took the path of least resistance, which is normally the channel of froth to the surface provided by the mother.

Another behavioral cycle begins when the little creature reaches the surface, or actually, any open space, for it turns out the new nymphs have small, club-shaped hairs all over that are under a membrane and kept flat against the body during digging. As soon as the nymph escapes into an open space, the hairs can stand up a little and, having a rounded club at the tip, they tend to push the membrane away from their young bodies without piercing it. This, I showed, is the signal for shedding the membrane.

Just as the little grasshopper will molt five times before becoming an adult, it must now, as its first activity in the wide world, molt the embryonic membrane. The process is similar. First, there is a long split down the middle of the back along a preexisting line of weakness, and the soft body inside swallows air, swells out through the split, and eventually escapes. It continues to swallow air because a lot of pressure is needed to unfold the many wrinkles in the still soft, thick cuticle of its first stage in the world. The head, which was so tiny and pointed for the journey up to the surface, expands dramatically, and its orientation becomes that of a typical grasshopper. In this newly emerged insect, the head becomes the largest part, with so many wrinkles and folds flattened. It is rounded out, and its unblinking eyes gaze, its antennae point forward. Its biting mouthparts test their movements.

We now have an insect that is obviously a miniature grasshopper, or locust, though it is still pale and greenish apart from those little black prehardened spots that enabled it to molt. It continues to swallow air, and all folds become completely flattened, all club hairs stand straight out, abdominal segments spread out, and one can almost see through the airy body. Air is held in the whole digestive tract while the plates of the soft, newly stretched cuticle harden and darken, a process controlled by a hormone that is released when those club hairs stand upright, indicating that the molt is complete. Few mechanical processes in nature are more strangely prehistoric than molting—a new body so cleverly emerging from an old skin.

Such was my PhD work. I loved every part of it. I marveled at the precision of all these complex activities, thrilled at the perfection that had evolved so many hundreds of millions of years ago, allowing insects to be among the most successful animals on this earth. And still, so many years later, I stop to watch the molting of a grasshopper or the emergence of a butterfly and remember all that I discovered years ago and how exciting it all was. I was finding things out, solving little mysteries, studying the wonders of nature, pondering the perfection that can evolve during the eons of evolution. And I had finally emerged myself into another new phase of my life. How privileged I felt. How lucky.

Palpating

Reg and I were fortunate. We were employed as partners at the Anti-Locust Research Centre in London, though we took care to be no more than colleagues in our work environment. One day I found him looking at photographs of the tiny feelers around the mouths of locusts. Each little feeler ends in a membranous dome covered with several hundred tiny pegs, invisible to the naked eye. The photographs had been taken using an electron microscope showing the pegs blown up to ten thousand times their natural size.

"Look," said Reg with urgency in his voice. "Just look at this, Lizzie. Some pegs have a round pore at the tip and others seem to have the pore closed to a slit."

"Oh," I replied, "I expect the processing for the microscope just caused some of the pegs to collapse shut."

But when I looked, I too felt surprise at a perhaps unknown phenomenon. Novelty always excites a researcher, and I said nothing for a few minutes. We both knew that the pegs were locust taste organs, and, being covered with hard cuticle, there would be a pore at the tip to allow chemicals to get inside. It was odd to find round open pores in some and closed slits in others, but the specimens had been freeze-dried, coated in gold, and examined under a vacuum—no wonder there were anomalies.

I looked at Reg puzzling over the pictures, with his familiar, plain face. My appreciation of him was not about his being good-looking, but such moments of puzzled interest and enthusiasm were definitely part of it. He looked up with a quizzical air that made me feel a rush of affection. He put away the photos of the palps, as they are called, but they were not to be forgotten—not by either of us.

As in so many puzzles, the pegs on the palps returned to consciousness from time to time, along with new ideas. They were known to contain taste-sensitive nerves, and Reg had already shown that locusts used the palps during feeding, palpating all over the surfaces of leaves, picking up chemicals that signaled *eat this* or *reject that*. Even the insoluble leaf waxes of leaves provide signals, transmitted to the nerves on special carrier proteins.

One day over lunch, Reg said, "Wouldn't it be interesting if the locusts could open and close the pores themselves?"

The seed was sown again in my mind, and as I walked to work or lay in the bath, I thought about the pores and how one could study those minute things in more detail. How exciting if taste buds opened and closed! How novel!

Reg and I had been working together on the ways locusts regulate how much they eat. What made them start feeding? What made them stop? Which plants were eaten and which rejected, and what were the cues? Our own mealtimes were often spent pondering such questions and talking about the possibilities. Knowing how feeding is regulated might have spin-offs for control of the pests, but for now, it was quite simply fascinating to figure out what made them tick when it came to eating.

Earlier, I had found that hungry locusts given young grass blades eat discrete meals, just as we do, and then stop more or less abruptly to take a long rest before once again becoming active and searching out food. The meal sizes on different types of grasses were rather similar, which suggested that they somehow measured a full stomach.

"Do you think there might be nerves that respond to stretch?" I asked Reg.

"Well, why don't you look?"

I opened up locusts and traced nerves to the digestive system. Then I began experiments to see if I could find particular nerves that signaled a full stomach. Each locust was opened up and one or other of the nerves to the stomach cut (or none were cut), then the insects were sealed up again. After the locusts recovered, they were allowed to feed, and all of them did. I compared meal sizes to see if the ones with a cut nerve ate more, indicating that this nerve was a pathway to signal fullness.

Eventually I found one small nerve that was key. If it alone was cut, the locusts ate much more than usual during a meal and kept trying and trying to eat more until their mouths were stuffed with food that physically could no longer be swallowed. I have a photo yet of a locust with chewed grass bursting out of its whole mouth area, like a little kid who wants to eat more cake. Clearly, this tiny nerve was the one that determined when the locusts should end a meal.

When I looked inside the locust with a powerful microscope, I found that the small nerve was connected to a pair of special glands near the brain.

"Maybe the nerve impulses to the glands are important," Reg suggested.

I thought about this. Maybe the nerve stimulated the glands, causing them to release a hormone that affected their behavior. The next step was obvious to me.

"I could make an extract of the glands and see if that affects feeding," I replied.

And that was my project for a few weeks. I was excited to find that injecting extracted hormone really did stop the locust from feeding. We were slowly solving a puzzle about how locusts feed, but it turned out to be more complex than we first thought.

It wasn't all about hormones. In humans and other animals, the body's nutrient levels are important in regulating feeding; we found that locusts ate more frequent meals if the protein level of their food was low. High levels of blood nutrients also reduced feeding. Plant chemicals of all kinds also influenced feeding; when a chemical tasted bad, locusts made smaller meals of foods containing those chemicals. Water content mattered, too. A thirsty insect ate more of a wet food; a well-hydrated insect ate more of a dry food.

As I worked on all these details, Reg and I studied the results and, at each new finding, wondered what needed to be done next and who might join the group to try and understand all the elements involved. Anthea examined the detailed structure of all the taste organs—scatterings of them on the mouth, feelers, and feet, including, of course, the pegs on the palps. Perhaps this would shed light on how they worked. Wally set up a recording device to measure how the taste organs detected particular chemicals. He discovered cells sensitive to sugar, salts, and amino acids in the pegs on the palps. Input from these would tell the locust to feed. Steve made continuous observations of locusts over days and discovered a rhythm in activity that tended to recur at twenty-minute intervals. Peter studied the taste organs of the locust feet and found that they responded to chemicals in the leaf surface.

"Great," said Reg. "It would be like people being able to determine milk or dark chocolate with their fingertips."

The locust was becoming a model for how insects control their food intake. We traveled to meetings and gave talks. We put all the elements together to make a story about the regulation of feeding in an insect. The fly had been the model up to this point—a slew of experiments by Vince Dethier and his students had made flies famous in the entomological world, added to which Vince was good at writing popular stories about his work. *To Know a Fly* is one of his most amusing books about how he and his group had made their discoveries. Now the locust was challenging the fly, and the story was not the same. Meanwhile, at the back of my mind, I continued to think of pores that might open and close at the tip of pegs on the palps.

I said to Reg, "We need to find a way to discover whether the pores might open when they are hungry and close when they are full. Wouldn't that be something?"

He agreed. "We need to examine them again with the scanning electron microscope."

We had another look, this time comparing palps on individuals that were hungry versus well fed. No pattern emerged of pores being open more in the hungry locusts, but in so many of the samples the whole dome of each palp had caved in that we were left once again with the feeling that processing them for scanning had changed the pore structure.

Other researchers dismissed the possibility of pores being able to open and close, which somehow made me hope it would turn out that they did.

One day I read a research paper about measurement of electrical resistance across biological membranes, including across insect cuticle. Suddenly it came to me in one of those "aha" moments: I could take live locusts and measure the electrical resistance across the palp tips containing those hundreds of taste pegs. Surely when the pores were mostly open, the resistance would be low, whereas when all or most of them were closed, the resistance would be high. I determined to compare electrical resistance on hungry versus well-fed locusts.

"Read this," I said to Reg.

He agreed with me about the plan, and then the question became how to get the equipment. We could find no source, so I went to a night class to learn how to make the electronic circuit that would be needed, and after a couple of weeks, we had the little gadget.

A tiny capillary with a salt solution was placed over the whole palp tip, and a wire was poked into the blood of the locust somewhere inside. With amplification of signals, I would be able to monitor the resistance. As I prepared for the big experiment, Reg and I talked about it.

"Wouldn't it be such fun if we found a difference?" I said with excitement.

Reg said, "Sure, it would, but you must do it blind."

In other words, I should make measurements on insects without knowing which individuals were well fed and which were hungry, so as to be completely sure I wouldn't unconsciously bias the results. Reg selected each insect and I fixed it carefully to a small stand, placing the capillary over one palp and inserting a wire gently through the cuticle at the base of the palp. I passed a tiny current through the preparation, first in one direction and then in the other, and measured the resistance. Then I measured the resistance across the patch of cuticle without any sensory pegs. At the end of the day, we pored over the data.

It was late into the night when the picture emerged, and we were so excited. "Let's open a bottle of wine!" Resistance *was* higher across the domes of the palps when the locusts were well fed than when they were hungry, whereas the cuticle with no pegs showed no change.

"Better repeat it," Reg warned. "You know people will find it hard to believe."

I made several runs of the experiment, and the results were consistent, leading us to spend evenings at the pub trying to think of alternative explanations as we drank our pints of warm bitter. We had to be quite sure. One night, Reg had an idea: "Let's see if Wally can use his electrophysiology setup to measure changes in single pegs."

Wally didn't need persuasion, and in the end, it was clear. When he tested the response of individual pegs to a stimulating solution, he only got a response from a few of the pegs when the insects were well fed, but from most of them when the insects were hungry. It was enough for us to feel that, yes, the pegs must be open or closed, and that feeding closed the pegs.

The three of us wrote the paper and sent it off to the *Journal of Insect Physiology*, whose editor, Howard Hinton, sent it back by return of post writing, "What a ridiculous idea—it can't possibly be true." We were incensed. We went out for a lunch together to discuss our next strategy and determined to send it instead to a more prestigious outlet, *The Journal of Experimental Biology*. None of us had published there, so it was thrilling when reviews came back: "exciting," "novel," "publish at once."

Palps were becoming famous, but we were not finished with them yet. Readers of our paper were fascinated but kept asking, How? We thought about where to go next. We thought about the hormone released when the insects fed and thought of Jenny, who had worked on the glands that release it. Jenny became our next collaborator.

Jenny and I sat down to dissect out the glands and make a hormone extract that could be injected into hungry locusts. The idea was to mimic the hormone release that occurs when locusts fill their guts. Once again, we took resistance measurements, and we were even more excited this time. Extracted hormone also caused an increase in resistance of the palp tips but not the plain cuticle.

Jenny had never really been convinced about our study, and I felt she worked with me partly out of politeness, but as the results came in she had her "aha" moment and smiled with delight. "Hey, Liz, it really works!"

Putting the facts all together: feeding filled the gut, which stimulated the little nerve, leading to release of hormone from the glands, and the circulating hormone caused closure of the pores at the tip of the palps as well as making the insects rest. We'd discerned one more factor involved in the regulation of feeding in locusts. It was just one of many factors, but it was something most unusual in the animal world. Who ever heard of taste organs shutting up shop so that food could not even be detected?

We never did find the precise mechanism that effected the pore closure, but the phenomenon of pore closure is now accepted, and when I meet other biologists at relevant meetings I get comments such as, "Oh, you and Reg were the ones that did that work on the palp-tip sensilla," or "Yes, I remember, opening and closing—*Journal of Experimental Biology*." For the rest of our working lives together, we both felt that this was one of the most interesting early pieces of work we did as a team.

That was long ago, but when I see grasshoppers out in nature here in Arizona, I can't help watching them on plants, waiting to see them feed, observing them palpate all over the leaf surfaces, and I imagine the tiny pegs on the tips of the palps with pores that can open and close to help regulate whether or not to feed. And I think back to the years of experiments, the wonder of discovery, and the satisfying life of working with so many people who were all excited about making each small step in discovering what controls feeding in locusts and grasshoppers. It remains to be seen whether synthetic chemicals could be sprayed on crops to close the pores of locust taste organs in order to limit destruction from the insects.

Rambling into Hungary

I had no reason in 1956 to learn anything about the history of Hungary, our school curriculum being about the British Empire and the Commonwealth, with Australia at the center. How would I know about the compelling complications of Hungary's birth, the Magyar invasion led by Árpád, devastation by Tatars, the Ottoman rule, the long history of mixed blood, the area's cultural importance in the Renaissance, its role in the history of Europe? How would I know about the Austro-Hungarian Empire, the comings and goings of Habsburgs and Communists, the support of Germany in World War II when a half million Hungarian Jews were killed, the appropriation of Hungary into Stalinist rule in 1948?

After the 1956 uprising, two hundred thousand refugees found homes in other countries, including Australia, whose government offered settlement assistance to around fourteen thousand of them—a significant number for a country of around nine million. Three years later, I had the fun of working during the summer at a giant cannery in the small town of Mooroopna, about a hundred miles north of Melbourne on the Goulburn River. Earning money for my next year at college, I worked the lucrative night shift. At five thirty in the evening, I showed up with my plastic apron, pushed my way towards the doors, and, along with a couple hundred other women, made my way to the rack where time cards were stacked in alphabetical order. I went through the turnstile, and, provided it was before six, the starting time of six o'clock was punched onto my card.

The remains of the day shift were still leaving, and a group of women wearing head scarves trailed behind until I heard shouts of, "Hurry up, get yer time cards stamped, you geezers." I found myself, as usual, on row two. I lined up on one side of a set of belts twenty yards long with six other women, while seven lined up on the other side. The cannery had eight such sets of belts, each beginning with two pear machines. The lucky ones, who got to a pear machine through excellence in other tasks, were able to increase their earnings, because working pear machines was piecework—the more pears fed into the machine, the more money that could be earned. But the pace was fierce, and to really earn more than the rest of us on hourly wages, you had to be in terrific form. I have no figure for the number of pears placed per minute, but it was a dramatic sight to see the racing hands, picking and placing, picking and placing. And the lurid cursing if there were a delay in supplies that would impact pay packets.

From the pear machines, half pears, peeled and cored, came tumbling along the outer moving belts at waist level. Between two such belts, a third one moved at a slower pace but a much higher level, and onto that we placed the perfect halves that had no skin or core remnants. This belt led to the canning section for export. We removed imperfections on the remaining halves with a small tool, then put the pieces into cans attached to a frame. Each worker had such a can, and when it was filled, the forelady emptied it onto a fourth belt, which carried the fixed-up pieces to the canning area for domestic sales.

Our job was called "specking." Cards attached to our apron strings at the back were clipped whenever the forelady emptied a can of specked pears. To keep the job, one needed to have at least twenty-four cans emptied per shift. Those who filled the most cans were rewarded with work on a pear machine when there was a dropout, and competition was quite intense, though the atmosphere remained jolly and friendly.

"You still at school are ya? Must be learnin' slow, eh?"

"Get on wiv it. Mavis done six cans already."

"Hey, hurry up mate, and ya have time to put a drawin' on the pear."

"A face or a willy, ha, ha."

"Just look at her bum in those red trousers sitting on that old box. Looks like a big red moon."

"There's the nine o'clock bell, wackadoodle. I'm ready for smoko."

At midnight, the bell rang again, and we got a half hour of rest on whatever box or bench was available. Men handed out lunch boxes and rolled tea trolleys up and down the rows. Women smoked and gossiped.

"Look at her."

"Ning-nong."

"Got a bun in the oven, I'd say."

"Yeah well, she's a good-lookin' sheila, and I seen her playing fast."

"Aw, half the blokes in the cookin' section wanta do a naughty with her."

By six in the morning, after twelve hours on the job, conversation flagged and everyone was ready for fresh air, breakfast, and sleep. Everyone, that is, but those women in scarves, who were slow to clock off, just like the ones in the morning. After a few nights, I realized why.

"The Hungarians wanna work more, ha, ha."

"Listen to them jabbering away in their lingo. They oughta talk English, I say."

"Yeah, new Australians. Look at them waiting as long as possible so as to get a few more bloody minutes on their clock."

"Miserable bloody Wogs."

"Alwis cadgin'."

Happy-go-lucky Australians had no idea what these refugees had been through, how every penny mattered, how they had reason enough to be serious. The job that was novel and fun for me was serious for the Hungarians.

I was not to know that other Hungarians would enter my work life properly later. I sailed to England and in my first teaching job in London met Rosalind, who taught history at the same school. We quickly became friends. Spring 1964 was bursting out when Rosalind offered me a place in her rented flat near Regent's Park because her friend was moving out. But first I had to be approved by the Hungarian landlady, Mrs. Kontuly, who was not happy with the way my ring finger had a bend at the last joint—and on both hands. She saw a lack of integrity. Even though she was satisfied with the long lifelines on my palm and thought the branched heart line was unusually good, the head line strong and curved, there was something crooked. It was just the finger.

"I don't like."

"You honest?"

"Where you living now, eh?"

"What you do for working?"

"Where you coming from?"

"You friends Rosalind?"

"I take a chance on you den, okay."

The flat was on the ground floor and basement, and I was to take over a small room at the back, next to the bathroom. Rosalind, as the longest renter, had the big front room with a bay window looking out onto wide Albany Street. Downstairs in the basement, Lorraine had the dark front room looking into the area. With a kitchen downstairs, and even a dining table, I felt I would be moving up many rungs in the scale of things, but best of all was the grand London street, with the famous Nash terraces and all of Regent's Park and spring glory to walk in.

No one knew anything about Mrs. Kontuly's background, except that she had escaped from something in Hungary. She never wanted to tell, and she never spoke of her husband, but she had the money to buy the lease on this fine Regency house. She had filled it with antique furniture, and, like other leaseholders in the row, she paid for the upkeep of the white-painted stucco front and columns along with the big black front door with its brass letterbox. Who she had been in Hungary I never knew. The stairs up to her domain were close to the front door, so we rarely glimpsed her mysterious friends who came and went, calling to her in Hungarian as they left.

It would be ten more years before I actually traveled to Hungary, when I was involved in my new career that made me feel as though I was being paid to do the things I loved. I had become an entomologist with a job in the British government and was speaking at an international insect-plant conference held at Tihany, outside Budapest, in 1974. Tibor Jermy, who was to become my dear adopted Uncle Tibor, hosted the meeting. It was at that meeting he grabbed the attention of entomologists from around the world and became my supporter.

Tibor was born in 1917 in what is now Slovakia. After he obtained a doctorate in zoology in 1942, his career was interrupted by Soviet internment. Here, he once told me, was where he learned to love languages. With other inmates, he learned and read German, Italian, Russian, and English, and he retained a love of languages; he was always happy to hear new English slang and to ponder the diversity of words in English. At his tiny apartment in Budapest, pride of place went to the dictionaries for translating four languages into Hungarian.

The meeting at which I met Tibor has many special memories for me. It was my first international scientific meeting, and the one where I first presented data that interested other entomologists. It was my first visit to Eastern Europe, behind the Iron Curtain. I conversed with people whose names I knew only from the scientific literature and began my life as a member of a friendly group of researchers scattered around the world who were fascinated with each other's findings, happy to get together at intervals to revitalize their togetherness, and thrilled to be part of a scientific family.

The meeting was held in a hotel at Tihany, a village on the northern shore of Lake Balaton. Tibor, then director of the Plant Protection Institute of the Hungarian Academy of Sciences, had organized our program and our entertainment. One evening we went to a concert at the nearby Benedictine abbey, founded in 1055, the later-built baroque façade of which dominated the village, and we looked out to the smooth lake and the sunset. On another evening, we ate goulash, drank far too much Tokay wine, and danced wildly in a cellar restaurant.

I have little memory of Tihany outside the conference hotel, however, so enraptured was I with the drama of being a scientist at an international meeting; the amazement of talking to Professor Müller from East Germany, who constantly looked over his shoulder and whispered of his son not being allowed to learn English; the sight of the Russian delegates, who were required always to be in pairs and separate from everyone else; the fun of my compatriot Alan standing on his head in the foyer of the hotel because he had hiccoughs, while the famous and very serious Oxford behaviorist John Kennedy managed a smile.

It was Tibor who gave me most at that 1974 meeting. His gentle warmth and quiet smile were there for everyone. His simple humor and genuine welcome made all of us feel wanted and created an atmosphere of friendly collaboration. With his thick gray hair, dark skin, and black mustache, he could easily have been taken for a Turk, and he had a mischievous smile that lit up when he told his jokes or made fun (just a little bit) of anyone pompous. It was a time in the history of plant-insect interactions study when new theories were developing that would attempt to generalize about evolutionary history, and a sense of excitement reverberated through the week. The field was not yet vast enough to prevent everyone being interested in all aspects, from the minutia of aphid responses to plant surface waxes and the chemistry of strange plants to the elaboration of grand evolutionary scenarios. Scientifically, the meeting was also Tibor's triumph. His talk, "Insect-Host Plant Relationship—Coevolution or Sequential Evolution," remained the hot topic, then and for years afterwards.

For a decade in the United States, the big picture of insect-plant interactions had been about coevolution, the insect-plant arms race, and the role of strange metabolites in plants as principal players in the evolution of herbivorous insect diversity and specialization. Tibor's argument was that plants radiated and developed chemical peculiarities for reasons unrelated to herbivory and that, later on, the insects had to adapt in various ways. In other words, there was a sequential set of events rather than a stepwise race between plants and insects to defend, adapt, and defend again. His theory was counter to the current dogma, just as a somewhat impish Tibor loved, and he had a case he could defend rather well.

The presentations took place during the week strictly in alphabetical order so that Tibor's talk came towards the middle, but it created great interest, intense argument, and, for me, much wondering. Like Tibor, I have a predilection for rocking the entomological boat, and I found the process of making a new scientific argument highly stimulating. Tibor celebrated discussion and fostered my addiction to new ways of thinking, different ways of looking at things at a time when there was a tendency to slavishly agree with the popular theory of coevolution.

The meeting in Hungary played a significant part in my scientific growth and marked the beginning of my lifelong friendship with entomologists round the world, most especially with Tibor Jermy. Four years later, Reg and I ran the next international insect-plant conference in England in 1978. Uncle Tibor came, of course, and loved my presentation that, similar to his in Tihany, rocked the boat.

Several biologists had theorized that tannins were the ultimate plant defense compounds, to which insect herbivores would never be able to adapt. The widespread occurrence of tannins in woody plants was seen as a generalized defense, preventing insects from digesting them and using their proteins; large concentrations of tannins were referred to as quantitative defenses, as opposed to the minute quantities of alkaloids and other toxins that were labeled qualitative defenses. It was thought that specialist insects could evolve metabolic pathways to deal with the latter but that the tannins were invincible.

My presentation demonstrated that a number of grasshoppers had no trouble dealing with tannins, and some of them even used the chemicals for tanning their own cuticle, thus poking a hole in the grand theory. My friendship with Tibor grew and flourished. His protégé, Árpád Szentesi, spent six months in my laboratory in London, and in 1982 in the Netherlands, the three of us presented a joint work on how experience changes behavior—another complication in under-standing how genetic changes evolve.

I met with Tibor and Árpád at the insect-plant conference in France in 1986, and then the meeting of 1989 was in Hungary once again. The Budapest conference center became an international scene, the field of endeavor enlarged, with the diversity of work encompassing all biological disciplines. Tibor and I gave related talks in the section on evolution, but by this time, we had somewhat differing views. Tibor elaborated on the theme of his 1974 exposition, while I argued that predators had an important role in the affiliation of herbivorous insects with their host plants. Our affection intact, Tibor and I argued warmly, then and for years afterwards, as we each saw flaws in the other's case, though now, the multitrophic theory is gaining precedence.

Even into retirement, Tibor and I wrote emails to one another through Árpád's computer (Tibor was unable to keep his, as it belonged to the

Hungarian Academy of Sciences). He sent me manuscripts to polish the English, and I sent my memoir essays to him. We shared a love of nature, wonderment at the brevity of life and the glory of a sunset. Across the world, we each listened to the music of Kodály, and just as Tibor remembered the open, friendly nature of Americans he met in his year in Washington State, as well as other visits to a free world, I remember my Hungarian friends and my visits to Hungary. I think of the many Shakespearean plays and concerts in production in Budapest when I visited in 1980 and the fine Verdi opera I attended for a mere three dollars, the hostel I stayed in that had once been a monastery.

I remember visiting Tibor's apartment, perhaps four hundred square feet, where his wife Grete made tea and we sat at a tiny table in the kitchen. I had brought, at his request, a chain for the door and a mask for Grete's eyes, as they had no blinds or curtains to keep out the morning sun in summer. He talked of the tragedy of 1956, of the Hungarians' disappointment in the West. We walked out along the narrow cobblestone streets, past medieval buildings restored after having been reduced to rubble in World War II. We walked over the chain bridge across the Danube, and I know that both of us felt an irony in the music of Johan Strauss floating up from one of the tourist boats.

The strongest memories of my two Hungarian friends find a combined love of science and classical music, a fascination with human behavior, and a melancholy for Hungary. Árpád, so ardent about visiting parliament in session in London, Tibor nostalgically bitter for what didn't happen in 1956, both of them pessimistic in what seemed a very Hungarian way. Both remained so, years after the collapse of Communism and the adoption of free multiparty elections. It makes me appreciate how years of political desperation and the disappointments of decades retain their power to shape our dispositions. Yet these two represent for me the excitement of my work career and make me remember the amazement with which I joined the fraternity of biologists from around the world.

Uncle Tibor and Árpád taught me so much and connected me with Hungary, their home, a country fraught with pain over centuries. I believe the burden of such a history plays out in the country's more recent problems and in the personality and character of its citizens—a pessimism combined with dark humor, a focus on world problems rather than self, and an old-fashioned idea of scientific honor. I think back to my student days, when I knew so little, and the Hungarians who worked hard in an Australian cannery with never a smile. We laughed at them, not knowing their sorrows or all they had left behind, not understanding how history makes us who we are.

Indian Medley

One morning I stood in my experimental field in southern India and looked across the patchwork of bright green sorghum and gray-green pigeon pea fields. A clump of trees marked the village of Patancheru, and nearby stood an old mosque with grass growing out of the dome. I had seen it for weeks, but that day I had a rush of love for the place—the scene, but also the history. During my first year of work in India, I would see crumbling temples of Buddhists and Jains; past and present edifices of Hindu or Muslim society; remnants of ancient cultures, wealthy Indian rajas, the British empire; and evidence of independence and partition. I would become more aware of India as a country with a long and very rich history, and as a place of extraordinary contrasts. I had learned Indian history in high school, at least the British period of it, read Forster's *A Passage to India*, and seen *Gandhi*, but here I really was!

I focused again on the fields. Rich, red soil with healthy plants in tidy rows—this could have been any agricultural research station, but the cart pulled by a Brahman ox, with its long horns and patches of pooja powder from Hindu religious ceremonies, was India. The jewels in this green and brown rural landscape were the multicolored lines of women in saris among the crops: weeding, cultivating, sowing seed. There was such a line in my field, spread out across the rows of seedling sorghum. Ankle bracelets and toe rings adorned their bare feet. Dark south Indian skins intensified the bright colors of their saris. It was amazing how easily they worked in such clothes, squatting all day long. Talking, they inched along as the boy supervisor stood watching dressed in flared trousers with a tight-fitting shirt. His hair was Brylcreemed. New sandals and a watch displayed his status. I consciously captured the details of the colorful, industrious scene for memory before rushing my plant samples to the lab.

Reg and I along with our colleague Sue were working on plant chemistry and insect behavior and helping to develop a sorghum variety resistant to the pest caterpillar *Chilo* that eats the plant base just at its growing point, causing it to die from dead heart. The young leaves wilt first, as rot sets in deep in the growing stem, then all the leaves wither and brown, until finally the stalk collapses. Looking along rows of a half-grown crop, the dead-heart plants can be counted with ease—sad, centrally collapsed specimens, half the height of their healthy companions with no chance of producing grain.

Our assignment as British aid scientists was to collaborate with Indian insect and plant biologists at the International Crops Research Institute for the Semi-Arid Tropics (ICRISAT) located outside Hyderabad. It was one of those projects that entomologists call "applied" or "problem solving" as opposed to "basic." For some academics, applied work is second best, there being less scope for major sweeps of ideas, major changes in paradigm. For me, though, the problem held plenty of interest and certainly offered room for original thinking and novel ideas—and I enjoyed the possibility that something useful might come of it, something that could eventually improve life for some of the world's poorest people. Grain sorghum is a staple food for people living in dry regions of a dozen countries in Asia and Africa. At ICRISAT, different teams focused on ways to make the crop more productive, more drought resistant, and less susceptible to pests and diseases.

As well as working in Patancheru, we visited towns and villages, temples and forts across India. I walked through ruins and markets and city streets. Everywhere I went, I was conscious of hordes of people, from low-caste street sweepers and laborers living in cardboard boxes to well-dressed professional men and women, rich businessmen, and film stars. And they traveled. Indian pilgrims filled the overcrowded buses and trains and gathered at shrines, holy places, and monuments. At the Taj Mahal and the Delhi Red Fort, I had expected a lot of European and American tourists, but the throngs of people I saw all seemed to be Indian.

At Patancheru, there was the same range of rich and poor, the same festivals, the same busyness. Much of the research concerned improving pigeon pea, chickpea, or sorghum, because ICRISAT holds the world germplasm collections for these crops—the seeds of all the possible varieties, from which one might find genes conferring resistance to pests and diseases or ability to withstand drought or salt in the soil. My colleagues were researching plant growth, grain size, pest resistance, and nutrition. Indian women regularly came to the station also, to make chapatis in order to test the culinary quality of the flours ground from grains of sorghum and millet and from pulses like chickpea and pigeon pea.

Our fields of sorghum were seeded with different varieties so that we could assess plant characteristics amenable to genetic improvement. We looked for increased resistance to some of the pests, including the caterpillar *Chilo* on which I was working. In the modern concrete buildings, insect rooms held cultures of healthy, unparasitized insects for experiments, and there were modern chemistry labs to analyze plant constituents. In a controlled-environment room, we tested the effects of plant chemicals extracted from sorghum varieties in which dead hearts were rare. Would one of these chemicals added to a synthetic diet and fed to caterpillars have some bad effects on them? Already, we had found two chemicals that reduced the growth rate of the larger caterpillars.

I worked early mornings in the field, so as the research progressed, we decided to relocate to the dormitory at the research station rather than make the twelve-mile daily journey from Hyderabad. On moving day, I felt almost nostalgic already for what I wouldn't see each day ahead: cars going each way on both sides of the street, cows with their humps and floppy dewlaps lying in the road, very slow oxcarts, rickshaws and bicycles everywhere. A thin old man in his dhoti pedaled furiously, sweat glistening on his bare ribs, as he pulled a bicycle rickshaw carrying eight little schoolgirls. Their uniforms were identical to those worn by little girls in England after World War II; I had an almost identical one when I was in grade school in Queensland, Australia. The girls sat in public school neatness, their plaits of black hair tied with white ribbons. Here, as in certain other details of life, time seemed to have stopped in 1947, at independence.

Stalls selling Ganeshas were set up in front of untidy fields of six-foot-high sorghum. It was the day to celebrate the elephant god Ganesh, my favorite of the Hindu deities, with a chubby body and an elephant head. He is the son of Shiva, the great god, and also the god of wisdom, the god of learning, and, at the same time, the bearer of success. On this day, Hindus should not look at the moon. On this day, Hindus buy plastic Ganeshas and, at the end of the festival, march in processions and deliver them to water tanks or reservoirs.

Approaching Patancheru, the familiar segments of huge concrete pipes lying around on the side of the road came into view. The new water line to Hyderabad was coming—sometime. Whole families had taken up residence in each segment of pipe, and round the entrance to the agricultural station all were overcrowded. The pipes were about three feet in diameter and ten feet long. Glancing down into one of them, a sea of dark faces looked right out to my car. The prime location, close to the gate where day labor was selected, presented additional benefits, such as gifts of old cups or tins, cigarette butts, broken umbrellas. On this day, many had been left unselected for work, and an aura of lethargy surrounded the pipes. Two men squatted smoking in front of one, their dhotis wound tight round their thighs; others stood leaning on pipe sides. A woman in an emerald-green sari leaned over a tin on a small fire, stirring, and several naked children sat in the dirt playing with sticks. Meanwhile, we glided in our imported car through the gate to another world, where World Bank money had financed high concrete walls and wide stairways, plate glass and stainless steel—spacious air-conditioned labs with the most modern equipment for biological research.

Sometimes I spent Saturday in Hyderabad with my colleague Sue, who was a plant chemist from England. Long legged and blond with an aquiline nose, Sue really stood out in India, yet like me had developed a strong affection for the place and loved to visit the city. The four-hundred-year-old city was a dense mass of humanity, a mix of Hindu and Muslim culture. Seas of bicycles and crowds of people swirled across Osmania Bridge and round the beautiful old buildings of the Muslim empire and later Portuguese and British states. We didn't dare drive around in this jostling, bustling metropolis of bicycles and rickshaws and animals and people, where crashes seemed always imminent. We couldn't bring ourselves to take a bicycle rickshaw, to have one of those thin men sweat and labor with our weight. Instead, we hailed a motor-rickshaw, pulled by a man on an underpowered motor scooter, and ignored the hundreds of seemingly near misses as our driver swerved to avoid boys on bicycles or cows calmly chewing in the road.

Markets tumbled into the streets: vegetable stalls around the mosque with unfamiliar species of plants, row upon row of aromatic spices in stalls around the Charminar, streets of shops selling silk. Puris and chapatis and innumerable unknown foods on trays and from cafeterias sent their mouthwatering aromas into the air, and above all were the mysterious smells of incense from stalls, shrines, and temples. In some places the stores were so tiny that they held just one man or woman sitting, with everything for sale within arm's reach; some little stores were actually built on top of others. Then, on the sidewalk, there were men doing shoe repairs with old car tires, cutting the hair of boys on stools, embroidering shirts with treadle sewing machines, and a woman selling pictures of the gods Parvati and Lakshmi.

In the narrowest lane, running with urine, sloppy with cowpats, we aimed for a jeweler's shop Sue knew. When we reached the house, we saw no indication of what would be inside. Up a dark stairway, past several furtive-looking men with shoeboxes under their arms, we reached the second floor and the jeweler. Customers sat on rugs, spreading out the precious stones on cushions; I had never seen so many jewels outside glass cases. Men in Nehru-style clothes stood at one end, and a group of rich Hindu women with rolls of fat bulging between the folds of their saris sat on one rug while a Muslim family sat on another. On our rug, Sue and I looked only at semiprecious stones, but the Indian men were patient. I selected a clear green stone, at the same time taking in the sight of newspaper parcels, all manner of old boxes, and even matchboxes appearing around me, each of them revealing more diamonds and opals—Hyderabad's unexpected treasures.

It was some months later, when I visited the nearby Golconda Fort and tombs, dating from the great sixteenth century Qutb state, that I discovered the area had long been famous for precious stones. It was near Golconda that the Koh-i-noor diamond had been discovered some six hundred years ago. Its early history is shrouded in myths, but it formed part of the loot of an Iranian shah who sacked Delhi in 1739, then it fell to his general, whose descendants were forced to surrender it to a Sikh ruler. In 1849, the great diamond was acquired by the British and placed among the crown jewels of Queen Victoria; today it is part of the royal crown of the British monarch, Queen Elizabeth.

Each of us from England made Indian friends at work. Reg and I received our first Indian invitation from Dr. Agrawal, another entomologist working on a different pest of sorghum, the shoot fly, that destroys developing grain. His was a tiny house of plaster-coated local brick near the busy high road. The bold blue paint was peeling, but a brand-new concrete image of the god Shiva stood by the door. We entered directly into a room that had space just to fit a bed and three small chairs. From this main room, we were led to a tiny kitchen with a table. He spoke Hindi with his wife, who served curry, but who stood behind us as we ate. After dinner, we sat in the bedroom chewing paan—rolled up betel leaf containing spices and who knows what drugs. As we chewed the paan (but felt no effects of drugs) we looked at our hosts' wedding photo album—a splendid tome with thick gray paper and a shiny, white plastic cover.

The pretty teenage wife in the photos from a rural village stands with her elderly parents. She wears a brightly colored sari, and ankle bracelets adorn her bare feet. Her eyes look down. The couple had not met beforehand, and her family probably had little to give as a dowry. Dr. Agrawal stands tall and proud beside his equally proud parents. He wears white cotton with strands of flowers round his neck and a wreath of them on his head. He is twenty years older, but looks gorgeous. He is marrying late. Perhaps there is a reason why his parents have only now made the match. Perhaps the astrological signs were perfect. There are many photos. The ceremony is the long, traditional Hindu wedding. I slip the paan into my purse when he looks over at his wife sitting on the bed. "I like the system," he says. "We don't have your divorce rate."

Dr. Gupta's apartment in Hyderabad was quite different, relatively spacious, with a television set. He told us the neighbors came to watch it there. From the painting on the wall, I guessed they followed Vishnu, the god with four arms, each one carrying a symbolic object. He is the creator and sits with his wife Lakshmi, a snake showing they are together for eternity. A bookcase contained the great Indian epics, including at least six books of the Ramayana.

Mrs. Gupta, a medical doctor trained in England, wore a magnificent blue and orange silk sari and golden sandals, and the mark of red dye on her forehead was set in a circle of gold. Both husband and wife had soft voices and an assured manner. Their demeanor told me that they were rich by Indian standards, traveled, educated—Brahmins, perhaps. A professional woman, at any rate, suggested a higher caste. But I was still feeling my way in a foreign culture. We all sat together for the meal, which was delicious and served with care. It was hot and highly spiced. We drank a lot of water—boiled, they assured us. We talked of Indian politics. What would happen to Mrs. Gandhi? What of the Sikhs? A servant took away the dishes, and we sat with our tea to look at their photo album—wedding pictures, of course, but also snapshots from Dr. Gupta's college years in Canada, where he obtained his PhD. It seemed that photo albums were obligate entertainments at Indian evenings, or was it just for us foreigners?

Meanwhile, I threw myself into the work. At the research station, I rose early, before sunrise. My first job was out in the sorghum fields. I had already learned that the mother moth laid her eggs at night in a scrinch of dead leaf at the bottom of the plant, and that the tiny new caterpillars must climb a meter or two to the top. Only at the top were the leaves soft enough for their minute jaws; only there could they survive. They experienced their Everest in the first light of day, before the hot sun would shrivel them, before the dry wind could blow them into dust, before the dew had gone. For weeks I monitored these specks of life, quantified nature, examined the survival of the fittest. I watched each hatchling emerge from its egg, then eat it and orient upwards. I saw it struggle through the wax on the stem, then crawl up the undersurface of a leaf only to discover it was not at the top of the plant. It would eventually find its way back down the long leaf to the stem and then continue upwards. Finally, after perhaps an hour, it would scuttle down into the depths of the whorl at the top of the plant to feed on a soft, new, developing leaf.

Why did the mother moth impose this dreadful climb on her off-spring, beset as it was with microforests of hairs, thick foot-clogging wax, crevasses, blind alleys, and tiny predatory enemies round corners? I found that if she put her eggs in a convenient place near the top of the plant, then parasites would find and destroy them. All the egg batches I placed on leaves up the plant became parasitized; only those at the base of the plant survived. Only bottom-laying females would have surviving off-spring. Thus, genes for egg laying at the base of the plant would be passed on, while genes for laying eggs elsewhere would have no future.

I focused on this magical early part of caterpillar life, a time when the young insect was most vulnerable to the hazards of weather and all manner of small predators, and eventually found that, on some resistant sorghum varieties, the tiny caterpillars were unsuccessful at the climb. They were able to climb well enough, but wandered out onto the under-surface of upright leaves and seemed unable to reorient, unable to return to the stem—an essential skill to have if the top of the plant were to be reached. Instead, they found themselves on the tip of the upright leaf, where they wandered about, then finally fell off or blew away. Why were they having trouble? I realized the answer to this question could be the key to the plant's resistance. I needed to discover the reason why baby caterpil-lars lost their sense of direction, their concentration, or whatever it was preventing their success. And so came the next step—experimentation.

First, I made replicas of stems and leaves. Fresh plant material was placed into a liquid that set quickly into a fine-grained rubbery substance, making a perfect mold of the original leaf and stem. I then cut into the mold, to expose and remove the plant material; into this mold I poured jewelers' resin containing green dye. The resin hardened in the rubbery mold, forming into the exact physical shape of the plant parts, right down to the small hairs and spines along the leaf edges.

My tests on caterpillars had to be done outdoors, at dawn, because this is when they hatch and climb. When the air is calm and moist, they climb towards the lightened sky, and they need to do it without delay to preserve their energy and moisture.

I lifted the newly hatched individuals, which are just a couple of millimeters long, by a thread of their own silk using a fine camel hair brush, and I allowed them to land at the base of the artificial plant standing in a tray of dark earth. It was a triumph to see the tiny caterpillars walk up the stem and out onto leaves as usual. But they didn't return to the stem as they must do on a suitable plant, meaning that there was indeed something lacking—was it the plant surface wax? Next, I applied waxes extracted from leaf surfaces, and here lay the fascinating details. With wax from normal, susceptible plants, caterpillars crawling up the undersurface of the leaf blade always turned back down when they hit the leaf edge. With wax from resistant plants, or wax from plants that were not hosts at all, caterpillars would reach the leaf edge as usual, but instead of turning once again to reach the stem, they carried on and out to the tip of the leaf, wavered and wandered, and eventually fell off or parachuted away on threads of silk. Clearly, the chemicals in wax could trick these larvae. If the wax didn't carry the right message to their tiny taste buds, they would hesitate instead of climbing up, they would remain on a leaf instead of reorienting up the stem. Then they would spin their threads and blow off to their certain demise on the morning breezes.

I was elated at my discovery. Lying on the red soil before the women went to work and before the men walked up and down watching over them, I recorded the travails of caterpillars, and had my personal revelation. The plant genes could be tweaked, not just to make those crevasses impassable and the wax stickier, but also to give wax the wrong message. With a different chemical in the wax, we could make those baby caterpillars toss their own lives away. Clearly, those chemicals had to be identified in order to help develop a crop with fewer dead hearts. It may seem strange that such tiny beings depend on the taste of wax to make the right choices for their lives, until one realizes that these vulnerable creatures crawl through mountains of wax, that they are surrounded by it for the first hours of life and have no chance to taste anything else. They are born with evolved taste buds specialized for detecting what they need to know and with behavior that enhances survival on the proper plant.

Sometimes an expatriate group went into Hyderabad for a meal in the evening, usually after one of us declared boredom with the daily rice and dhal provided at ICRISAT's cafeteria. Our favorite was the Emerald, tucked away in a side street, but bright with green fluorescent lights. The dishes were the best of Indian foods that any of us had tasted—masalas, vindaloos, and bhajis—perhaps a choice of fifty mouth-watering foods and, always, fresh lime soda to drink. Most of the Indians ate with their fingers, having mastered a technique of rolling up the rice with whatever else in two fingers and a thumb and miraculously conveying each parcel to their open mouths without spilling any. Instead of napkins, one went to a row of washbasins and rinsed off as needed. Some foods were served on banana leaves, and, in some cases, one's plate was a banana leaf. I had lived in London long enough to try out and enjoy some of the hundreds of Indian restaurants there, but none of them compared with the Emerald for quality and atmosphere.

About a week after my wax discoveries, I was working in the lab with Sue to make wax extracts from leaves for chemical analysis. Only then could we move to the genetics. We dipped the leaves in different solvents, separated fractions, purified precipitates, and prepared samples for identification in either the gas chromatograph or the high-performance liquid chromatograph. Chandran and Situ worked with us. Both men loved to speak in English and had so many questions about America and, especially, England. We finished the job early that day, and Situ took the glassware to Aurora, who stood in her pink sari at the sink washing glassware all day long. She was lucky because her job was permanent, unlike the jobs of the field laborers whose work was a daily lottery and depended on season. After Situ had gone, Chandran shyly asked me and Sue to his place for tea. A short, stocky, yet very handsome man, he said very little and appeared unconfident, being in a more junior position in the institute, but we had developed a rapport by working together. Mostly he spoke in Hindi or Urdu to the women who did the majority of the actual work in the caterpillar-rearing rooms. With his gentle smile and his soft dark eyes, Chandran was passionately Indian and passionately Hindu, and we were thrilled at the invitation.

At Chandran's house, we sat in a circle with his wife and children in what seemed the only room. Newspaper cuttings covered the grimy walls, all of them pictures of Gandhi or sayings of Gandhi, speeches by Gandhi. We ate bananas and Indian pastries and special candies—too sweet, but we felt we must accept them. Each one had a layer of silver paper inside and I asked, "Do we eat all of it?" The answer was, yes, the silver paper was good for the intestine. The children were allowed just one each, but as visitors, we needed to eat several in order to be polite. The water was certainly not boiled, but just as certainly, we felt we should drink it with the rest of them.

Conversation came back to Gandhi time and again. How much he had given as a spiritual leader as well as a political one, how much India was for Indians, how much the new India had achieved. I felt the tension of a lower caste couple, the diminished confidence, the intensity of nationalism, their pride in a new India. There was the Mahatma on the salt march, the Mahatma in prison, the Mahatma on hunger strike. Chandran's wife looked tired and worn, but she smiled with obvious affection at her husband's intensity, and my heart went out to her. We could not communicate, as she spoke only Hindi, but I felt conscious of her personal pride and knew there was a distance between us, perhaps greater than language, that was not to be easily bridged.

The work became routine and tedious. We tested extracts and partial extracts. We tested groups of chemicals and single chemicals. And always we watched the tiny creatures and what they did at the edges of artificial leaves. But our motivation was the thrill of discovery forthcoming, the excitement of solving a problem, the possibility of preventing dead heart sometime in the future. We labored on. Sometimes Chandran helped us, but most of the Indian scientists preferred chemistry or biochemistry or genetics to lying on the dirt out in a sorghum field, especially after rain.

Our work ended after a year, though the result was not quite as simple as we had hoped. The wax profile mattered more than any single chemical—the proportion of various wax components was responsible for the differences in neonate caterpillar behavior, and the genes controlling these proportions would take time to find. Different levels of gene expression eventually turned out to be the answer, and still other genes controlled this in turn. The need to solve this problem would drive this work in other labs and other countries. Pursuit of the control of the chemistry of wax production would progress down other paths for other reasons also—needing to reduce water loss, wanting to mess up the behavior of greenfly, trying to prevent the entry of fungal diseases. Discovery of the precise genes involved would be delayed until the genome of *Arabidopsis* (a tiny plant, and the first to be so well understood) was worked out, which happened after I left India in 1982.

The Indians would continue where we left off, and they would find additional mechanisms of resistance, other vulnerabilities of *Chilo*. I went home to England with India in my heart. For months, I thought of Patancheru and wanted to live in that wonder-mix of a country, amongst the friendly, hardworking colleagues there, amongst the smells of incense and spices, the colorful religious celebrations, the love of animals, the extraordinarily rich culture. I wanted to remain where I had that revelation with caterpillars at a big research station. I wanted to continue down the path to finding the genetic code of sorghum and the end of dead hearts.

India remains a glorious memory. When I look at my photos— Reg with Indian friends at the Red Fort in Delhi, Sue eating street food in Hyderabad, temples built into the cliffs at Ajanta, Indians holding on to the outside of trains packed to bursting, oxen chewing cud in the middle of a busy street, a giant Shiva carving on Elephanta Island—my love of India overwhelms me all over again. When I visit Indian grocery stores in London or New York or San Francisco, the smell of incense and spices and the sight of tea or custard powder in old-fashioned British-style packets bring on a wave of intense nostalgia, memories of the great edifices of a complex history and multilayered culture, multitudes of people working at the institute, and hordes of people everywhere. Even more intense is the memory of tiny caterpillars in the early mornings when dew was on the leaves and mist hovered over the old mosque at Patancheru. It was a time when I fell in love with life and work and all things Indian, but also a time when I became aware that, although I had multiple nationalities, I would always be an outsider in that country of riches and contrasts.

Color in the Tropics

"Crikey, not again!"

It was usually Bill who suffered and I who turned to laugh as he pulled down his pants and brushed the army ants off his legs. But he had the last laugh this time, as I suddenly discovered even larger numbers of the creatures had found me and I had to jump away, similarly undress, and go through the frantic ritual of removing them from my hot, sweaty body, extracting them from folds of damp clothing.

However carefully we worked along the rows of the cassava crop, there were many occasions when one of us simply didn't notice the long columns of maddening, marching workers, great waves of them approaching, turning even the crickets and frogs that jumped up in front of them into prey. The Nigerians were more alert apparently, because I can't remember that Sam or Tolu suffered from them.

Of all the dreadful pests and diseases I had learned could be a part of the tropical West African experience—malaria, Guinea worm, sleeping sickness, diarrhea, Buruli ulcer, to name just a few—it was only army ants that ever bothered me. The species, *Dorylus wilverthi*, can have colonies of twenty million workers, sterile female fiends that swarm out in fans looking for prey on a twelve-meter-wide front, usually at night. But in the early mornings, remnants of the night raids and smaller bands of the busy creatures are just as voracious.

Our jobs at the United Kingdom's Centre for Overseas Pest Research were governed by the requirement to work some of the time on controlling pests in developing countries, and the projects had to be agreed upon administratively (politically). Each overseas project was set up in partnership with the developing country, with scientists from their institution and ours jointly tackling specific pest problems. The UK provided most of the money, and the recipient country was expected to help with accommodations and transport. The Nigeria project we were working on ran through the early 1970s. Reg and I traveled there four times, each time for four months.

The south of Nigeria where we mostly worked is hot and very humid, with a relatively short dry season. Agricultural areas are cut out of the rainforest, and when the soil loses its fertility, that area is abandoned to forest regrowth and new patches of forest felled. Many farmers grow a mixture of crops: lots of cassava, also called manioc or tapioca, together with corn, papaya, okra, tomato, and peppers. We were there to work on *Zonocerus variegatus*, a two-inch-long grasshopper brightly colored with red, yellow, and orange on a shiny black background. Historically, it had been a fairly uncommon insect, but in the middle decades of the twentieth century, it had reached pest proportions and each year completely defoliated the cassava crop in southern Nigeria. The questions were: What had caused the change in pest status? And what could be done to control populations?

Reg was in charge of the project, and I was one of several other entomologists directed into this work. Bill was there continuously for a five-year period and settled with his wife in an expatriate-populated area of Ibadan, where they had a cook, a cleaner, a gardener, and a night watchman. Tall, thin, and pale with a black beard, Bill stood out in Ibadan. He was ideal for the job of keeping the field records and looking after the routine data collection year-round because he was so mild mannered and patient. He ensured that those of us who came for shorter periods had all we needed to get straight to work in the field the day after we arrived.

It was 1973, and I had flown from England to Nigeria with Reg for the first time. As I walked down the steep steps from a Nigeria Airways plane onto the runway in Lagos, a new and overpowering combination of smells greeted me: wet, rotting tropical vegetation; untreated sewage; too-sweet floral scents; sweating African bodies. I had been in the tropics before, but here, the hot, humid midnight air was so oppressive that even breathing seemed a labor.

We followed the crowd across the tarmac to the single-story terminal, where temperatures were higher even than those outside, in spite of fans. Planes from Europe all arrived in the middle of the night, so the place was milling with people, including many Nigerians, all with sweat glistening, beading, dripping, making their clothes dark under the arms and in the middle of their backs, running down their legs and making sandals slippery.

We joined the long queue for passport control and were accosted almost immediately by men in Nigerian dress offering their services.

"Sir, Madam, I take you to the front, one naira."

"Come sir, I take you to the desk, one naira only."

I was reluctant to give money away so soon or to give my passport to strangers. How did these men come to be on this side of passport control? We watched as offers went to everyone. A man would grab a willing passenger's passport and urge, "Come, come quickly," and then he would rush to the front of the line, where arguments ensued with all those up there crowding around the desk.

Hundreds of airport employees walked about. Line-jump helpers rushed up and down urging others to come with them, hoping to earn lots of naira this night. Women swathed in yards of brightly colored cloth with fantastic headgear of the same material screamed as they called out to children or friends; the babies cried. Money changers called out black-market prices in Nigerian naira for English pounds and American dollars.

An hour passed, and we were further away from entry than when we first arrived. Eventually we were last in line. Reg remained silent. At the desk, when we finally got there, the Nigerian official glowered. He took my passport and leafed slowly through. When he got to the photo, he turned it right way up, looked at me disdainfully, and waived me through. I waited to see if Reg was treated differently, but the official glared at me. "Go." I went.

Now we had to go through customs. What is all this film? Packs were opened and exposed to see if they had been used. What is this net? Roars of laughter at the idea that we should run after insects. The batteries we brought for equipment were confiscated. I could feel moisture from my forehead running down the side of my face. Our video equipment was the greatest problem. It was tempting to invent a wild story, because it required patient explanation to make them believe we were in Nigeria to study an insect pest. We eventually succeeded. Or the officials became bored.

Just as I turned to Reg to smile my relief, a group of laughing young men wearing only shorts grabbed our bags. "We are from Interpol. Routine check." It was so patently unlikely, but there we were, strangers in a foreign country. How very dark they were, and how pallid and sickly we must have seemed, the only Europeans in sight; lank, brown hair stuck to a wet, pink neck contrasted with jaunty, tight black curls on shiny black skin. And they were so full of life. They went through our bags with wild hilarity; I didn't get the jokes, but I laughed anyway. I began to pass into a lightheaded state where nothing mattered in my funny, crazy, hot dream. Then, finally, they were done, and we went through the doors into the thick, black night.

The draft of air coming in the cab window brought me to life as we sped screeching along, splashing through puddles, veering round car wrecks, bumping over who knows what, passing food stalls still open along the roadside with crowds of people eating, drinking, laughing. Electric lights glared brightly, and I read the neatly made billboards above the low buildings: car doctor, tailor best shirts made here, one-day watch doctor, quick gonorrhea cure. The "cure" was in small type, so the sign was misleading at first, and we both smiled. At a crossroads, I saw the body of a man lying in the middle of the street. The cars and mammy wagons swerved to avoid it, but no one stopped. We went past the "barbing saloon," where a panel of painted faces showed different hairstyles: the rounded effect, or flat on top and sides, or curved out at the sides and then flat on top, like topiary done on English hedges.

We were aching to rest when the cab turned into a driveway. The lobby of the Airport Hotel, though it had a slight moldy smell, was agreeably cool and filled with the roar and clatter of air conditioners.

"Not long now," I said.

But we were not there yet. "No rooms left," they told us.

Ten minutes later, the receptionist commanded, "Stand over there." We did so, and we stood for some time. In my head, I said all the buggers and fucks. Later, we were told, "Go to room 716." This was a command we liked. We sat on our beds and laughed as we drank bottles of STAR, the Nigerian beer. No drink could be so good as this one was.

In the morning, the taxi ride back to Lagos airport gave me a taste of the months ahead, when we would work in the fields early each day. It had rained in the night, and steam rose in the first heat of the sun, creating a misty morning. I expected things to start early, but the streets were quiet, and the stalls that had been so lively the night before were dead. I saw a sleeping body under one and a few bare bottoms squatted over the foot-deep storm drain along the side of the road. The body at the crossroads was still there.

Our plane was ready, but there were no seat assignments on the seventy-mile flight to Ibadan. A noisy crowd suggested the flight would be full. Bodies lounged across chairs or jumped up and down in their cotton clothes, brilliant colors on black bodies. Reg and I were both quite refreshed, and yet, we felt pale and lifeless by comparison. They shouted their jokes and demands and suddenly burst into laughter.

The flight announcement came, and there was a jubilant rush for the door, but before we got out onto the runway, a woman in a green uniform instructed us to pick up boarding passes from a heap of them on a table. Reg and I reached for the bits of green cardboard, and I picked up two by mistake. It didn't matter, I quickly saw, because a second uniformed woman at the bottom of the steps to the plane happily accepted them both.

An hour passed. There was no air conditioning, and the plane became unbearable. It had been oppressive outside, but inside the plane became so hot and fetid that even the lively Nigerians went quiet. Finally, we transferred to another plane and took off at about lunchtime. Clouds were gathering as they often did in the afternoons, and I knew that if planes arriving in Ibadan ran into cloud, they turned back to Lagos. We were lucky—clouds didn't close in. We flew low enough to see the rainforest below and clearings here and there with small villages and their few crops. Most of the passengers seemed to be from the populous Yoruba tribe that lived below, and they were madly joyful.

Ibadan came into view as a sea of mud houses with rusty corrugated-iron roofs almost engulfed by tropical forest. This city, which was the largest in Africa south of the Sahara at that time, had between one and three million inhabitants; no one knew the exact number. Few Europeans lived in or visited Ibadan, and it was not a tourist destination. Here, in 1973, tall buildings were rare and the narrow streets unpaved. Storm drains provided the only sewage system, and the vast rainfall washed it all away to somewhere else.

We could see the grassy plain of Ibadan airport below a patch of cloud. The wheels were lowered with grinding and squeaking, then the plane descended fast. Suddenly, perhaps when we were only a hundred feet from the ground, the plane jerked sickeningly and we rose again. I saw animals on the runway; sheep, or maybe goats, were all over. We circled three times before we finally descended again, landing with a bump. I was more than pleased to experience this jolt of arrival.

Looking back now, I remember so much detail about Nigeria—all the new, exciting markets and shops, the noisy colorful bustle, the smell of street food, and the sound of treadle sewing machines—and, of course, our work. But most of all, I remember my first twenty-four hours—the hot, humid night, body smells, and masses of black humanity. Two sickly-pale foreign people among so many thousands of relaxed, comfortable, laughing people of rich, dark color in all in their glorious, colorful cloths.

At 8:00 a.m., Bill, Andy, and I worked along the rows of cassavas counting *Zonocerus* grasshoppers—live ones and dead bodies. Tall, quiet Bill with his gentle smile had all that we needed—the tally counters and record sheets. He moved with long, slow steps and spotted the insects quickly. Andy struggled. He was delicate looking, with his slight body and lank, white hair. It was his first field trip, and he hesitated over every detail, bent over to check every leaf, and was easily distracted.

"Just look at this jewel beetle. I must take it home for Jenny." He was in Nigeria without his wife for three months.

"Oh god, I trod in a turd," he said upon encountering his first bit of human excrement.

"What in the name of fortune . . . ?" He pointed to a gourd hanging in a regrowing rainforest tree.

"Oh yeah," Bill replied, "you see them all over. They have frogs and all sorts of things in them. Black magic."

Mystery emanated from the magical gourd in the earthy moisture of the morning. I looked inside and saw only black liquid with indeterminate lumps emerging on the surface. We would never know the meaning, but the materials rotting in there made up a hanging counterpart to the rapid disintegration of dead leaves, rotten papaya fruit, moldy insects, and human excrement, the fast reduction of complexity to component molecules. What invocation, I wondered, accompanied its making, and what hopes or curses were suspended here. But the black mystery was not for ex-colonialists to understand.

The hoppers were easy to count because they were so conspicuous and not at all shy. The dead ones, suffering from fungus infections, always climbed to the top of the plant and expired in the position best suited for dispersal of the fungal spores. Being gregarious, live individuals tended to cluster around the dead bodies, which is the best place to be infected themselves. Reg was the record keeper, walking beside us with the clipboard. Every day we counted all the party-colored grasshoppers eating the cassava crop; their red, yellow, and black seemed so very appropriate in the tropical heat. We were always soaked with sweat as we walked along the rows, and we often had to avoid the great swaths of army ants that stormed through the lush vegetation, massed onto frogs and crickets, and rushed up trouser legs with such abandon.

We collected specimens of the grasshoppers to dissect in the laboratory for parasites. We monitored their movement in the field by labeling individuals with radioactive phosphorus and then searching for them daily with a Geiger counter. We observed parasitic flies attacking mating pairs. We tested over one hundred different plants for palatability to the grasshoppers and spent weeks studying the ability of different plant species to support growth and survival of their young. We worked hard together, and we laughed as we thought of new words to describe what we saw.

"Hole in neck," called Andy as he checked for exit holes of parasites.

"F 27," called Bill, the femur length measured.

"Four maggots," I added as I dissected a fat female.

For years after, we smiled as we remembered "hole in neck" and all the fun of working together in the sweaty heat of the humid tropics. We sat together in a London pub retelling stories of army ants and sodden plots; of team work in a tiny lab where so much equipment didn't work; of wild parties at night in a Lebanese restaurant; of shopping down in the market, where a hundred voices loudly urged us to buy bananas, fried plantain, mangos, coconuts, chilies, okra, fly-covered meat, multi-colored cloth, wooden carvings, and strings of beads. Young men offered to make embroidered shirts then and there in half an hour; women held up squawking chickens for sale or shouted for us to examine the bush meat—rats, usually.

Tony was our official Nigerian scientific collaborator. He was a lecturer at the University of Ibadan, where we had our base, but we rarely saw him. He was busy with more important things. He was tall and unusually shy and quiet for a man of the Yoruba tribe. He directed his two graduate students, Tolu and Sam, to work with us.

Sam was from the Igbo tribe in eastern Nigeria. He was small and wiry, very smart and very polite. He took in all the details of the project and had interesting questions.

"Do you think," he asked, "that the grasshoppers may be more abundant just because the cassava is more widely grown now?" An interesting possibility.

Tolu was a Yoruba and very different from Sam. He was intelligent, but it was hard to get him to talk seriously. He was large with a wide mouth, and when he laughed, which seemed to be at almost every-thing, he rolled his head around, eyes glistening with fun. He came out in the field with us only a few times and couldn't contain his amuse-ment at Andy; there was something about his almost girl-like, tentative movements and something about his difficulty with understanding the Nigerian accent. Tolu had many jokes to tell. "You hear about man loving banana so much? He sees monkey up in the trees eating banana and says, Hey, monkey, I want to be monkey, so I eat banana all day." At which he laughed uncontrollably. It was such fun to watch him laughing that we all laughed, too.

After the first weeks, we rarely saw Sam or Tolu. I don't think they cared for the dirty, sweaty stuff of fieldwork. All the Ibadan students dressed in smart clothes, what looked like Sunday best, and I wondered how they managed to look so spruce when some of them, at least, lived in real poverty. We were messy by comparison, traipsing around in muddy pants and boots. For us, the casual, sweaty life was a welcome change from dressed-up London's high-heels-and-necktie workplace, but I wondered if our untidiness was an affront to the ironed Nigerians.

What did a team of British scientists find out in those five years, and what was the point? In the end, it was a successful project that spawned new biological questions and led to some interesting research findings. One that I was particularly fond of concerned the love-hate relationship of the rather gregarious insects with their cassava hosts. Individual adults moved about and regrouped on particular cassava plants, but the young preferred weeds, and even adults found cassava distasteful unless it was wilted. Discovering that turgid plants gave off too much cyanide to be acceptable, while wilted plants gave off little cyanide when bitten, making them highly acceptable and nutritious, solved the paradox. Herein lay one of the benefits of being gregarious: one or two bites from each of two dozen grasshoppers aided the wilting process and turned bad food into good.

But the principal story was about the pest, its status and its control. No particular plant species seemed important or necessary for the grasshopper's growth and survival. Fungal disease killed the greatest number of insects, but fungus infestations followed rain and could not be easily manipulated. And while parasitic flies killed off most adults before they could lay more than one batch of eggs, one batch of eggs per female was still enough to produce a devastating pest population the following season.

One morning, taking a short break out on one of the plots, Andy suddenly said, "Hey, just look!" Reg, Bill, and I saw that a group of Queen butterflies was hovering over one of the dead grasshoppers. One of them landed on its body, tapped it with little feet, then uncurled a proboscis and began feeding from liquid oozing from its anus.

"Danaids!" shouted Reg in excitement. "Alkaloids!"

Each of us knew the stories of danaid butterflies licking plants containing the pyrrolizidine alkaloids they needed to make aphrodisiac pheromones; this was the first clue that our grasshopper might feed on plants containing them. The butterflies were perhaps finding a source of the alkaloids in the liquefied gut juices exuding from the dead insect.

Zonocerus turned out to be a chemically defended insect that sequestered various toxic compounds from its host plants, including pyrrolizidine alkaloids from the flowers of *Eupatorium*. This introduced plant had become a weed over the previous fifty years, and in doing so had become a readily available plant with the required chemicals for the grasshoppers. The alkaloids were also broken down in the insect, becoming a volatile chemical important in attracting insects to one another, which resulted in great crowds of the pests congregated at egg-laying sites—all the females in an area of about two acres laying eggs in close proximity.

It was Bill who noticed the egg-laying crowds. "You guys have missed a big scene," he said one afternoon. He was dripping with sweat and sipping his lightly salted water and as excited as such a passive person could be. We normally worked early in the field and spent the heat of the day in the lab. But Bill had gone out at midday to check on something and found that this is when it all happens—masses of females laying eggs in a small patch of dirt, with males on adjacent bushes jumping on any female that goes by. From then on, we also worked through the heat of the day, at least sometimes.

At the end of the season, when the insects were all gone, Reg said, "Let's scrape the surface soil away at one of the egg-laying sites to expose the tube of froth that extends up from each egg mass." We all got on hands and knees with trowels. At each place where we found eggs, we inserted a match so that, in the finish, we could count the matches and photograph the little wooden forest. There were seven hundred and ninety-three!

Such aggregations had benefits for the insects. In particular, they led to dense crowds of newly hatched, chemically defended individuals, drastically reducing mortality from predators. Plenty of alkaloids, available in the new, weedy plant, gave us a possible clue to the recent increase in grasshopper populations.

With Wester, a Sri Lankan lecturer at the University of Ibadan, we delved into the question of how tolerant the insects were of the cyanide in cassava leaves. And Sam lapped up the guidance and advice that Reg was only too happy to give. For the most part, though, we worked as just a British team—several of us during the grasshopper season and only Bill in the off-season.

Controlling the weed that contained pyrrolizidine alkaloids seemed out of the question; it was everywhere. On the other hand, discovery of the vast egg beds led to a simple and successful means of grasshopper control—learn where the crowds congregated by going out in the fields at midday at the appropriate time of year, and then, in the fallow season, dig up the eggs, exposing them to sun and desiccation. This meant digging just a few inches deeper than normal tilling in the places where egg beds had been identified. We tried it out on several large plots of two hectares each and found that the pest populations could be reduced next season by 95 percent compared with the plots that were left untouched. The key was to be out in the fields at the right time, when the egg laying happens. It was such an easy method to regulate the pest numbers, and the technique became key to solving the problem. Bill stayed in Nigeria an extra year to provide training in this easy method of regulating *Zonocerus*.

What do I remember best now? Heat and mud in the field, where colorful grasshoppers congregated, butterflies flocked, and army ants marched; noise and color in the markets; the sound of revival songs from a church on campus—"Jay, ee, ess, you, ess, Jesus, halleeluyaaa"; constant pressure from men and women hawking woven or embroidered materials, indigo-blue adire cloth, wood carvings, food, masks, brass bowls. "Best for you, only twenty naira," the same ones called every day, laughing and slowly reducing the price on successive days. "Come on, sir, yes, sir, bargain, ma'am, today at only twelve naira." Always bargaining, laughing, and joking, it was hard to know whether to them life was just funny or if we were most amusing.

As for the project, I loved the teamwork and laughter, fascinating biology, and solving a pest problem. Grant proposals, authorship, and recognition were in no one's mind, but getting a job done together, satisfying curiosity, finding answers, and enjoying the novelty of life in humid tropical Africa felt like a life fulfilled. Nigeria is remembered by the English team for all the novelty of an African community built in rainforest and for working with Reg. Of course, he was not just our leader—he was my lover and life partner—but the team seemed happy with both him and us. Our team's work together was always harmonious and fun.

I also recall how quickly I came to feel different, how my pale skin became odd, how conscious I was of any slight, even if unintended. I think I learned something of what it felt like to be a minority individual. I wonder now if even the simple proportion of individuals of different appearance could be enough to make a majority confident, a minority less so. The biological precedent is strong and not just among humans; the unusual has often been considered peculiar, if not downright flawed—worthy of persecution, worthy of death. There is the history, the killing of Native Americans, of Tasmanians, of mainland Australians. We might imagine that modern *Homo sapiens* did it to *Homo neanderthalensis*.

At some distant future time, maybe we will all understand that our various appearances and habits and the relative abundances of different colored skins are the tapestry of our ever-expanding civilization, and that our individual threads must weave together to make an acceptable aggregate, a product that will last. My trips to Ibadan enlarged my appreciation of many things besides the biology and control of *Zonocerus variegatus*.

Swinging in London

Thelma wore a light, but well-fitting, scarlet sweater with a low neckline, making the most of those lovely breasts. Her leather miniskirt consisted of a dozen panels, each coming to a point at the hem, which must have been only an inch or so below her panties; it looked like a cross between a pixie outfit and a little girl's party skirt. Her perfect legs walking on high heels produced a hip movement that made her skirt sway seductively, and she had a habit of swinging her arms in a way that seemed she wanted to embrace someone.

Thelma had arrived in 1970 to be the new government administrator at our Anti-Locust Research Centre—soon to become the Centre for Overseas Pest Research—in London. She was a flyaway blonde with a model's body and an easy laugh, and, with it all, great sex appeal— the combination of hips swaying just a bit, shapely breasts shown off just enough, and eyes that, twinkling, caught the eyes of others. She wore the latest fashions, but nothing too extreme, and exuded the aura of a woman who knew everything she needed to know about her own attractiveness and its effect, especially on men. And we all liked her—men and women. She was evenhanded and thoughtful, efficient and smart. Best of all, she had an ability to get around rules to suit the sometimes-unusual needs of scientists who were readily frustrated by government regulations.

As usual, on this day she was at work early, dealing with the mighty piles of paper, including requests from staff and new requirements sent down from higher administration. She had wandered down the corridor to meet with John, who managed all the ordering of supplies for the scientific staff, a man who had to be checked at all times to ensure his wine bottle was hidden and his mental ability not too impaired. She poked her head into a couple of research labs to say hello and spent some time in the big secretarial room, where seven women and two men typed manuscripts and letters from handwritten copy. Large Lexie, draped in a loose, striped gown, managed the rows and rows of files, each bound with a faded pink ribbon—the proverbial red tape. Normally, Thelma could call for someone to bring her the necessary file, but from time to time she dropped in to the office herself and chatted amiably to the little army, especially Lexie, with whom she was particularly friendly.

It was a Monday, and senior scientist Edward Trumble, a craggily handsome man with a great deal of charm, came to work later than usual, his train delayed on the long commute into town. Everyone noticed and wondered if it was something more than just the train. Reg and I knew he was torn between trying to keep a marriage together and being completely infatuated with our Thelma. At work, we all knew that something was going on, but so much was going on at our institute that the presumed affair of Edward and Thelma had gathered little attention. Until that Monday.

The sixties had just ended, psychedelic, heady times of Pink Floyd and The Beatles, avant-garde theater and underground newspapers, willful hedonism and wild new fashions. A youthful, sexually daring counterculture expanded. Antiestablishment permissiveness pervaded even nerdy labs and could be found in the dusty offices of civil service organizations like ours. The English writer Duncan Fallowell once remarked about the sixties, "There was a lot of wacky stuff going on and people just used to fall into bed with each other all the time." It was the time of "swinging London."

Thelma went into Ed's office and heard him out about staying with his wife. Then, for about ten minutes, she swung up and down the long corridor past offices and labs with arms flying, face flushed, and hair disheveled, calling out at the top of her voice, "I am *going* to get a fuck. I *want* a fuck. You are going to *fuck* me. Fuck. Fuck. Fuck."

Work stopped, conversations died, and faces peeped through doors. In my lab, I looked over at Sue, who was busy making plant extracts that I was to test for their effects on the feeding behavior of locusts. "My word," was all she said as she looked at me with wide eyes.

I looked up and down the corridor and saw a few gaping faces, a couple of doors shutting, and then there was silence. It was almost too surprising to discuss.

From those near the scene of action, we learned that the two of them ended up locked in Edward's office for over an hour. None of us knew where the affair was heading, but afterwards both of them carried on their jobs effectively. And there were several of us with very close bonds ourselves who were sympathetic to their compulsive attraction to one another.

Sue completed the extracts, and we set to work putting them on wheat flour wafers that would be presented to individual locusts in small cages. The wafers were actually communion wafers, which turned out to be excellent food for the insects. We could compare the amounts eaten on wafers with and without particular plant extracts. We discussed the idea that had been the start of this work, that the locust species that specialized on grasses made their choice because of bad-tasting chemicals in all nongrasses.

Sue said, "You would think there would be something special about grass that made them eat it."

"Grasses are bland," I replied, "and nothing extracted from them makes *Locusta migratoria* feed on neutral substrates."

We avoided discussing Thelma's outburst, though when Sue was ready to go out to John's office to order more solvents, she first looked carefully out into the corridor, then turned back to me and giggled before stepping out on her mission.

The institute where Reg and I worked for thirteen years after I obtained my PhD seemed to have a high proportion of staff whose sex lives came to work with them. Apart from Thelma and Edward, Paul was the most conspicuous. He was a tall, lean man in his forties, with a tanned face that was both worn and handsome, a long, thin nose, and a smile that was friendly and wide, inviting and charming. At any moment, he might imply that you and only you could possibly be attractive to him. His loud, infectious laugh was for you alone.

He came into my office one day on some pretext and said something about sex that was outrageous but funny and something of a turn-on, though I don't remember now what it was. And he must have known exactly what he was doing. His hand was down my pants so suddenly I was overwhelmed for just a minute before I slapped his licentious arm and shouted, "Quit it, you sex maniac!" He laughed loudly as he strode out and along the corridor. In retrospect, it is remarkable how all of us women accepted the fact that some men were just like that and that you had to have your wits about you. There was no #MeToo movement back then.

He was simply adept at taking women by surprise. One learned to be prepared, but my guess is that quite a few women were carried away by it in his office, and certainly some became regular visitors. There were times when I knocked on his office door, then tried to turn the handle, but found the door locked. I could hear the certain sounds of sex in progress and left them to it.

One of Paul's jobs was to edit all outgoing scientific papers in our institute. He was good at it, but the author needed to go through the paper with him in his office, to sit beside him and discuss points of detail. I quickly learned that his door needed to be kept open to the busy corridor, so when he moved to shut it, I jumped up and opened it wide. It is interesting now to think how audacious he was, how silent all the women were, except to one another.

"Oh, it's just Paul."

"Just make sure you keep the door open."

"He's just oversexed, you have to keep an eye on his hand."

The institute was a British government research organization funded with aid money, but not at all the stodgy kind of place I had expected. Apart from Paul, there was the dark-haired little librarian, Greg, who brought his gay lover to the library on weekends and managed sex among the stacks. "We are getting married today," he said when I surprised them.

There was Andrew, who kissed and fondled Judy on the stairs at lunchtimes; Helen, who openly slept with all the men who were interested "to help them"; Gwen, who had sex parties on Saturday nights;

and Hartley, who was rumored to have been a convicted sex offender. Of all the staff, there were fewer than 20 percent in a conventional, monogamous relationship, and of those, some were on second or third attempts.

Once when Reg was away in Pakistan and Edward Trumble was visiting the Philippines, Thelma invited me to a party at her house. She lived with a policeman called Mick, and many officers were there, boasting of the money they made in bribes. Lexie was there, too, with her enormous black lover, the two of them drunk and raucous and funny. Thelma trapped me in a quiet corner to talk about Edward. "Oh, how I want him!" she moaned.

One day, Reg and I were in a dark observation room, monitoring feeding behavior of locusts. Through observation holes in a wooden screen, the insects were well lit, and we were able to time the start and end of each meal. It was tedious work really, but at the time it seemed fun, and, anyway, I loved doing any of the experiments with Reg, who came out with ideas from time to time as we watched. Edward came to talk to us, and in the dim light we could see that his handsome forty-year-old face had lost its usual gaiety.

"What in the hell am I going to do? How can I leave Margaret, even though it is a pointless marriage? What would she do about the big house? How will my difficult teenage son deal with a separation?"

We could only listen. What could we possibly say to Ed? It was so hard to make judgments for someone else.

Some months after Thelma's famous parade of fucks, we were given another display; this time, she came to work in a low-cut blouse and a shocking-pink silk miniskirt. Her rant this day was not as vehement, but it was again full of anger and fucks. It created less of a stir than the first time, but seemed to increase restlessness among the staff, along with more gossip of possible sexual encounters. Enid, the sixty-year-old spinster, was said to be having an affair with the head of another department, and there were stories coming home of a wild relationship developing in India between two of our staff working on a project there. Brian was doing it with Vicki in the lab, while Paul was French-kissing Julie in some corner.

Paul's new routine was to get a woman close to him somewhere and discuss his own sexual activities, such as the need for his dick to say "Hi" to his current partner, Celeste, at many intervals during the day and night or how he and a few other colleagues (which ones, we wondered) had a holiday in France where they all swapped wives. This talk, it seemed, turned on one or two of his victims.

As the sex storm between Thelma and Ed rose to a crescendo, Reg and I went off to Nigeria to work on an agricultural pest and were joined by one of our young scientists, Andy. He had a wife and baby in London and had to leave them behind for three months. Such separations, I suspect, added to the broken relationships so common at our institute. Andy was new to the tropics, a slight, blond Englishman who worried over mosquitoes and ants, heat stroke and strange food. In Ibadan, he met a delightful girl working for USAID, the United States Agency for International Development, and was quite soon captured by her love of Africa, her comfort in a life of discomfort, and the ease with which she dealt with Nigerian ways. He fell for her, and the affair eventually ended his marriage—just another one of the changed relationships we were seeing in our work. Reg and I were lucky because we usually traveled and worked together, and our close relationship was everywhere known to be intense but stable and not really worthy of gossip.

Back in London, it seemed that nothing much had changed. There had been another Thelma fuck-walk, and Paul was still putting his hands down blouses. But within a couple of weeks, Thelma invited some friends, including Reg and me, to her new flat in Kingston. "You will get a surprise," she told us. Edward greeted us at the door and hugged us all. Then, Thelma hugged us, too. "We did it," they said in unison. And we could only give them our best wishes, hoping they would be as lucky as we were. We drank to that. And they did stay very happily together. Our experiences of broken marriages and subsequent passions were part of our friendship with them, a friendship that remained intact across the Atlantic Ocean.

At the time, I found it all amazing. I had no idea whether our government research institute was typical, and I have been gone too long now to know if such open sex in London is still routine and ordinary. I belong now to the United States, where sex is more secret and where there is also a less forgiving attitude towards anything considered improper. But my memory of those years working for the government is a kaleidoscope of hard work, research on pests, fun side projects, the opportunity to work with Reg, and, of course, the blatant sex lives of scientists at work. I'll never forget Thelma marching along, shouting, "Fuck, fuck, fuck!

Desert Time

1

I take up the trowel and a roll of toilet paper and head out into the sand dunes in the blinding midday sun. We are near the abandoned village of Tin Aouker in the Tilemsi valley, about thirty miles north of Gao in Mali (N 16048′, E 0008′), camping with tents, trucks, and Land Rovers. Reg and I have a tent where we keep our personal stuff, but we sleep on camp cots covered with nets under the stars. The radar team and the meteorologists must go to bed in their tents at sunrise and sleep during the morning hours, and they have recently risen to join the rest of us for lunch.

We have all been sitting in the mess tent—a high canvas twenty feet long without walls. Twelve of us are eating salty soup, fresh pan bread, and dried fruit. I wear a knee-length skirt, cotton blouse, and sandals. The men wear shorts with sandals or boots; half of them wear shirts. Ibrahim, our Malian cook in long baggy pants, is often by the opening to the kitchen tent, where a forty-four-gallon drum of water stands, supplying our needs for three days.

We talk of the morning's findings. Coulibali found hundreds of grasshoppers in the light trap, a four-meter-by-four-meter ditch we had dug that was lined with plastic, filled with soapy water to drown the catches, and illuminated at night with a lamp from the generator. Most of them were insignificant little brown things called *Oedaleus senegalensis*. Don and Mark had dissected some of them in the lab tent to estimate their state of maturity. Nick had searched the area for species that might be living in the minimal vegetation. Reg and I had taken a Land Rover further afield to see what we could find where the radar had suggested takeoff just after sunset on the previous evening. Everyone feels sleepy now in the heat.

I need to traverse about a hundred yards of lifeless sand before I am out of sight of the camp. I seem to be the only living thing in the shimmering heat, the temperature well above 40°C. I encounter the straggling growth of a small, bitter gourd, *Citrullus colocynthis*, and one grass plant, *Cenchrus*, with dried-up leaves and spiny seeds. But nothing moves. It is the time of day in the desert when it makes sense to be completely still.

No sooner is my action beginning when I see the thing coming and marvel at the way it flies in the heat, wonder where it could possibly have been hiding out of the sun. It is a dung beetle—a stocky, black, two-centimeter-long tank of a thing that comes slowly my way in a zigzag flight upwind. It comes at an angle, and when it seems to be past the proper path to reach me, it turns about ninety degrees and comes again. The movement is exactly what entomologists call anemotaxis.

Insects have a simple mechanism for getting to the source of a good smell, a smell that might mean food. The antennae do the smelling, but once the good quality is recognized, they automatically turn into the wind at an angle that regularly alternates a little bit, resulting in zigzag flight. This tends to keep them within the odor plume. If they reach the edge of the odor plume on the right or left, they deviate again to the left or right, and, thus, they reach the odor source.

I am thrilled to see such a perfect demonstration of the process, so I wait with my trowel, not wanting to spoil the dung beetle's reward, as I try to recall the evolutionary history of insects. Beetles have been around for hundreds of millions of years, but dung beetles? I don't know, but certainly for tens of millions of years. She lands right on the prize and immediately begins the process of getting it under the sand into the right sizes and positions for laying her eggs. After the burial, which takes no more than twenty minutes, I am more than ready for my own return to the shelter of the mess tent, to the all-important water, and to Reg, who will enjoy my story. I hadn't needed the trowel after all.

2

Six of us, five men and a woman, sit around a wooden table covered with green plastic cloth, each with a Fanta, the sickly sweet orange soda I have seen all over Africa. No one wants any of the dark-colored food in the glass cabinet, which is unidentifiable, though some of it is probably goat meat. A door with a torn screen blows open and shut. The beautiful boy of about thirteen behind the counter wears khaki shorts and a tattered white shirt. He stares at us, and I know he has George's attention. Such a boy will return to London with him later. It has happened before. Silent George, with his love of North Africa and of Muslim boys, is a key player in most of our locust and grasshopper projects. He is just one of the diverse characters I work with in this place where time has a different meaning.

Through dusty glass, I see two bearded men in long white robes squatting in the heat under a neem tree. One draws shapes in the dirt with a stick, his robe folded up over his head. The other gazes into the distance, cloth drawn tightly round his thighs. They talk little, and I wonder how they manage to squat for so long without moving.

I am in Gao, Mali, October 1978, as the only female member of a British-sponsored interdisciplinary team studying night migration of pest grasshoppers. We flew to Niamey in Niger, and for two days we have been traveling north, crossing into Mali, along a rough road that follows the Niger River upstream. We left some of our group at Daoga and will head north from Gao, away from the river and up the wide, stony Tilemsi valley. The two teams will camp fifty miles apart across the Intertropical Convergence Zone, an area of low pressure that forms where the northeast trade winds meet the southeast trade winds near the earth's equator. As these winds converge, they force moist air upward. This causes water vapor to condense, or be squeezed out, as the air cools and rises, resulting in rainfall. In desert areas, the Intertropical Convergence Zone, which moves north in summer and south in winter, is a place where grasshoppers may find plants, at least sometimes, some years.

Our food for the month was purchased in England and shipped to Niamey. Our water drums are filled in Gao, and we must get used to managing in camp with little more than a gallon per day per person. Each of us will develop a different strategy for using his quota efficiently, and mine, apart from drinking, is to use it for cleaning teeth with no more than a tablespoon, washing my face with two pints, and reusing that to wash the rest of my body or my hair. What is left of this I reuse again for washing clothes or my feet.

Here in the hot afternoon café, we daren't drink the water but instead drink that warm, very fizzy, very brilliant-orange-colored Fanta. It is good to sit still after rattling along for two days in a Land Rover, but conversation lags. Joe, who is fluent in French, talks briefly with the waiter. There was a sandstorm two days ago, drought continues, the best well has become contaminated with grasshopper corpses.

We hear the midafternoon call to prayers from the nearby mosque, and I realize that we have been in the café for a long time. Three men go by, one in long red robes, the other two in white. They prod a single zebu bull ahead of them. "Fulani," murmurs George, but they have gone by before I have time to register their faces. That I miss some details is unimportant. That I have no watch doesn't seem to matter. That time passes so idly seems reasonable.

Reg and I are silent. We are partners. I know he thinks, as I do, of the sights of the past two days, the villages of Ayoru, Tillabéri, and Ansongo, with their low mud houses and narrow streets full of people and animals in the cool of the morning, classic barren desert beside the wide, brown river Niger, large decorated boats being loaded with grain to be paddled or motored downstream to bigger towns. There is a wide alluvial strip between barren desert and water that is almost luminous green with rice, turning the adjacent red sand dunes into a delusion.

Two men on camels pass by in the street. They are Tuareg nomads, draped from head to foot in dark blue cloth. Only the top halves of their faces are visible under turbans, faces also blue from the dye of the cloth. I learn that, for centuries, these people herded camels and goats across the Saharan plains, leading a hard but independent existence. When the French colonized the region in the late 1800s, the Tuareg put up a fierce but unsuccessful resistance. Then, with independence in the 1950s, the Tuareg were parceled out among the newly created countries, and their nomadic existence restricted. Here, they sit erect and silent on their camels, and we will see more of them in the camp at Tin Aouker.

The sun is low. Dust-laden rays of sunlight hit the edge of the table-cloth, highlighting the feeding and copulating of flies. I notice that one of the flies has only five legs but is in no way hindered in its activities. If anything, it seems to get more matings than the others. As I watch them, I discover that they vary in size, in shades of gray, in speed of running, in how high they raise their legs, in how quickly they engage in coupling. They differ in how low they hang their heads when extending proboscises with spongy tips and in how long the sponges engage in licking old food spots on the plastic. Flies have been so engaged for almost one hundred million years, and I become conscious of time or, perhaps, timelessness here, so far from everything I know.

3

When we arrived at our campsite at Tin Aouker in the Tilemsi valley, it was almost dark. Reg put up our two-man tent by starlight while I struggled with our camp cots and nets outside the tent. We all got our water rations and turned in early, exhausted from the rough journey. The hot day turned into a cool night, and we needed blankets out under the stars, where the donkeys startled the midnight hours with braying, and we woke early with the dawn.

Don made his little camp away from the rest of us, saying, "I need distance." When he rose next morning, he found his cot beside a skeleton emerging from the desert sand, and it wasn't long before several Tuareg men arrived, gesturing and shouting. We had accidentally camped on an old burial ground with some special significance. The skeleton was to be dug up and reburied. We had to move camp. Don was one of those people to whom things happen. The next night, the nomads brought him back after he lost his way when he went off in the dunes to pee.

At the new campsite, Nick picked up a stone arrowhead, and we all marveled at the quality of the point, the details of the tiny notches. His blue eyes sparked as they often did. "Let's make one. Let's try hunting with one."

Then George found a stone axe-head. In the next hour or so, we all had found stone tools and we forgot grasshoppers in our excitement. Small boys from the nomad camp nearby came to watch us, and we gave them coins for any that they found for us. The tools were scattered all around on the stony slopes of a low hill, exposed, I imagine, as sand had been blown from the surface. None of us thought about the archeological value of the site, with tools in such abundance, and it was twenty years before Mali established laws restricting export of such items; I still have my collection in a leather basket on a table by my front door in Tucson, Arizona.

My arrowheads are between one and two inches long. Some are narrow, with very fine lateral serrations and sharp points. At the base, they have hafts that must have fitted into wooden handles, so that they look like very symmetrical miniature pine trees. One is broad, with five large serrations along each edge. Another has rounded sides and a particularly sharp point. The roughest looking has sharp, smooth sides and is almost completely flat. Each, I suppose, must have had a different use. The fine "axe-heads" were probably scrapers and handheld, varying from about an inch long and even smaller in width to about two inches square, polished and rounded with perfect edges. There are no signs of chopping or flaking to make the edge, making them Neolithic, at the earliest probably between four and six thousand years old.

I hold a scraper. It fits exactly in my palm. Each of my fingers runs over the silky-flat surface, as I imagine a dark hand long ago that may have treasured such a tool. I think about the hands that held the scraper then and what thoughts may have engaged that man in long hours spent grinding and polishing. Was there joy in making and holding this thing? Was it the Swiss Army knife of those times? And why was there no sign of wear on its sharp edge? Had he dropped it out of his woven bag as he ran from some enemy? Was it kept for prestige? Had it been used in a burial? Did the maker wonder about the meaning of human life? What tribal identity, myth, and ritual gave order to his days? What have we inherited from those distant times in our needs for aesthetic detail, quality tools, mythical explanations of who we are and why we are here?

4

George in Daoga was on the two-way radio talking to Nick up with us in Tin Aouker. "Our radar shows big swarms of grasshoppers heading north at five hundred meters, going at fourteen meters per second. Wind from the south at five meters per second. Over."

"When should we see them on the radar up here? Over."

"If conditions remain stable you should see them in about four hours, so let's say 1:00 a.m. We will need the information from your upward-looking machine, too, to get the species. Over and out."

Those of us with daytime jobs were sitting around the table under the mess canopy, chatting about the day's work. The temperature had finally fallen to something comfortable as we drank tea and picked at broken shortbread from a big square tin.

"Interesting that all the females are immature. No sign of eggs in ovarioles," said Don, who had been dissecting insects under the microscope in the lab tent.

Reg replied, "Well, that would be typical, not just for grasshoppers, but for most migrating insect species. They need to get where they are going before putting on weight." He continued, "What was the species profile from last night's light trap?"

"Oh, a mixture. *Oedaleus*, mostly."

"Yes, that's what I got on the transects around here today," said Nick.

"Liz and I went north to where the radar showed concentrated takeoff last night, and at first we saw nothing. Then Liz poked a stick down into the deep cracks left by drying mud in a wadi and, you wouldn't believe it, but it was full of resting grasshoppers, all of them *Aiolopus*."

As darkness fell, the radar team was busy in the truck, monitoring activity in the air above on an oscilloscope, and using a movie camera to record the changes on screen. Insects appeared as white dots, and a group of them at similar altitude showed as a circle of white dots. Groups at greater altitude created circles with larger diameters. With many insects in layers, the screen became a mass of concentric circles, but when the air was dense with them, the screen was entirely white. Dense in this context would be around four insects in one hundred meters cubed.

Our mission was to understand the nighttime migration of pest grasshoppers that were adding to the desertification of the Sahel in West Africa. Life for the nomads is hard enough in such a desert without the added problem of competition from millions of these grass-devouring insects. They were inconspicuous and cryptic in the daytime, but with conventional scanning radar, we could see them at night, and measure their density and flight speed. With upward-looking radar, we could detect the orientation and wingbeat frequency of the insects, which was used to characterize each species. We obtained additional evidence to identify species from ground surveys in the daytime and from collections of insects that dropped down into the light trap at night.

Meteorologists provided measurements of wind speed and direction at different elevations. The hypothesis was that the grasshoppers were programmed to fly downwind and thus pitch up at the Intertropical Convergence Zone, the best place to find food. And this proved to be so. There were several grasshopper species involved, and the masses of individuals flying at night were mixtures of species all doing the same thing. They flew before they were reproductively mature, presumably banking on the assumption of getting food and then settling down to eat, mate, and develop eggs.

Our team of people worked well together—radar scientists, electronics technicians, meteorologists, ecologists, insect physiologists, and Nick, the taxonomist and trickster. There were fourteen of us, plus the Malian locust-control man, Coulibali, and our Malian cook, Ibrahim. At sunset, with the day jobs done and the night jobs yet to begin, we all sat down together for a single, small beer and reported the highlights of the day: grasshopper survey counts, stone tool finds, Tuareg activities.

After the first few days at Tin Aouker, a few Tuareg women lost their reticence and gathered close to the camp. It was me they looked at, the unusual sight of a white woman in skirt and sandals. One morning, as I went to my tent in pajamas to wash and dress, I found three of them waiting just inside. I smiled a little nervously, not knowing how best to regain my privacy, and they looked at each other, giggling behind their blue robes. I noticed silver objects sewn to the bottom of one of the robes. I saw dark mysterious eyes in beautiful young faces looking up and down my body. I washed my face and hesitated, but they remained motionless. Eventually, however, I shooed them off, and they fled, laughing, back to their encampment.

One day, two Tuareg men came to our camp and explained in French that they needed water. We had brought the minimum, used it very sparingly, and certainly couldn't get into the business of supplying the nomads, but it seemed that they had an old man dying. Nick felt there was no option and filled a goatskin for them. It turned out to be a true story—the old man died next day.

Surveys sometimes took us close to the Tuareg camp. We could hear radios, with occasional French words emerging from crackle. Radios are typical gifts from Europeans or Americans crossing the Sahara after they run out of water and are saved by the nomads. When the batteries run low, no one seems to mind that hearing the words or music is impossible; the volume is kept up just the same.

Men in blue cloth sit in a circle under the scant shade of a solitary acacia tree, talking. Women in blue cloth grind spiny seeds of *Cenchrus* grass at the entrance to a tent. Children ride off on donkeys to a distant well, join us in racing after grasshoppers, play complex games in the sand. The sores on their scalps speak of scurvy. Darker-skinned Malian boys wearing only briefs appear to be treated as a subclass, carrying out many of the daily jobs. Camels laze. In the distance, a block of window-less rooms remains, built by the French in a vain attempt to provide a permanent township. In the other direction, heat mirages quiver over the stony hills.

5

We see it coming. A distant brown blur across the rocks and dunes of the desert. And the wind is rising. "Quick," yells Nick, "Get the tents down, chaps!" He is excited by the drama, but practiced at dealing with all emergencies of the camp.

We set to as the brown mass comes towards us. We flatten every tent over its contents and hold it down with rocks or whatever heavy objects we can find. We do the mess tent last, which is difficult with the wind blowing hard and the first sand blowing in our faces. John and Joe cover the radar dishes and tightly close the equipment truck containing the generator and oscilloscopes. Then we all get into the Land Rovers, roll up the windows, lock the doors, and wait.

For three hours, we can see almost nothing but brown as the sand and dust whip across the desert, sanding the paint on anything left exposed, scratching the glass on the windscreen. The vehicles rock. The edges of tents flap free from their anchors. The sounds are eerie with whines, and none of us speak. We listen and watch the violence, and wonder how long. I wonder how the Tuareg manage sandstorms and what on earth the grasshoppers do. Surely, the sand would wear away the waterproofing of their cuticles.

The air calms down and the sand settles. The storm is over by late afternoon, and we are able to put things in order again before dark. And that is good, because sunset involves a ritual that every single one of us values above all other things in this Mali camp. The generator, brought to run electronic equipment for the nighttime radar work, runs a small refrigerator during the day. It is just big enough to hold one small bottle of beer for each person, and every evening at sunset, we empty the day's ration with such smiles and goodwill that I am sure—after a long day of piercing sunshine, shade temperatures above 40°C, a monotony of warm water, salty soup, and hot tea at predetermined intervals—that beer is critically important to the success and enjoyment of the trip. I start to think of that treat quite early in the day and by midafternoon can think of little else. The faces at sunset when Nick calls, "Okay, chaps," tell me I am not alone.

The day of the sandstorm happens to be Sunday, and Don, who had thought out the whole menu for all these people in the desert for weeks and weeks, had decided that Sundays should involve some kind of treat. His choices overall had been excellent. We had muesli with milk made up from powder for breakfast, together with lots of tea. For lunch, we had soup made up from powder, some canned meat or sardines, Ibrahim's pan bread, canned or dried fruit, and lots of tea. Dinner varied. Sometimes we had fresh food from Gao, but usually we ate stew from dried meat, scrambled eggs made from powder, rice dishes from packets, cheese, Jell-O, and lots of tea. On Sundays, a big can of Dundee fruitcake was opened, and the fruitcakes marked the passing weeks. I never felt like eating in the still-hot evening, but all of us ate slices, if only to assure Don that he had done well. And so, on the day of the sandstorm, after sunset beer and simple dinner, we sit around the trestle table in the mess tent with lamps burning eating fruitcake, telling tales of life in the desert. From time to time, we hear the light splash of an insect hitting the water at the light trap.

6

I live now in the Sonoran Desert in southern Arizona, but by the standards of the Sahara it is hardly a desert, with its winter and summer rain and rich flora. But it is a desert, with its piercing heat and bone-dry wind and the strange expansion of time as we wait for rain, as cactus plants wrinkle, as shrubs lose their leaves or turn completely brown. An ancient arrow piercing my largest old saguaro is evidence of the people who once lived simple lives here, wondering, surely, about who they were and from whence they came, waiting out the dry seasons that still seem endless even with our sophistications.

Where we worked in Mali, there were just three plant species, and those scarce and restricted to wadis, the spots that are slightly lower, where water collects a little when it rains. It was not always so. Neolithic humans lived there in abundance. But it is the nature of deserts to make us aware of time, of the past, and of endurance. It is the nature of deserts to trigger contemplation of who we are and where we came from.

Under the vaulted dome of skies over a nameless continent, unhurried human existence passed its days there thousands of years ago and for thousands of years. What did time mean for those people? The stone tools in such abundance that must have taken long days to perfect suggest a rich living. So long after, we went by plane for a matter of weeks to that barren place littered with the artifacts of lives long gone and empty air where we imagined conversations, thoughts, and hopes of people who shared our genes. We left again by plane, taking with us insect specimens in bottles and on pins, information collected in files and notebooks, celluloid film recording the places and events, and exquisitely durable prehistoric artifacts—and with them, dreams of who we were before our world of electronic overload, reverberations of a past in which grinding patience and unrecorded ideas belied the future wilderness of a world enthralled with its own momentum and blind to risks of its own demise.

Growing with Pollards

My large kitchen looks out onto a courtyard of green palms, shrubs, and herbaceous perennials. On one wall is a poster of a painting by Vincent van Gogh. It shows pollarded willows at sunset, the sun's low rays filling the sky and shining out at me over tall yellow grass and between the blue trunks of the gnarled trees, most of them leafless. It is a well-known painting—there are prints and reproductions of it available from hundreds of websites. Van Gogh paintings have a remarkable texture that is, of course, missing in my picture, but the reproduced painting is startling nonetheless. No other picture in my house emits such excitement and radiance.

Pollarding, a kind of pruning that is less frequently used today, was evidently common round Arles where Van Gogh painted. It encourages a close, rounded head of branches, and puts the foliage out of reach of deer. The custom was refined in formal, manicured royal gardens of Europe for centuries, where the idea was to grow a durable tree to a fixed size and then maintain that tree at that same size forever.

The appeal of pollarding has waned, but the practice is still carried out on old trees where it has been performed for years and where the particular garden or landscape would be much altered by leaving the trees to grow without pruning. Somehow, pollarded plane trees represented my life as I learned to become an American, a professor at the University of California, Berkeley campus, where there are many such pruned trees. On my way home to a house on the hill above campus, I often walked through Sproul Plaza, where Polka Dot Man lay with legs in the air, Orange Man walked in his green overalls as he carried paper oranges to tie onto trees, and older professors dreamed of the free-speech movement in the sixties, still believing it was the only free-speech campus in the world.

I think also of pollards seen from the windows of Wellman Hall, the home of the Entomology Department, of which I was a member. One looked down into a square with pollarded planes in two rows, their knobby fists signaling upwards on winter mornings and their leafy heads shivering in summer breezes. Often I gazed out at them as I marveled at the strange fate of becoming a professor there. Wellman Hall holds my memory of so much in Berkeley when I was a newly appointed professor, straight from a job as a scientist in the British Foreign Office. It was my first academic appointment, my first experience of the United States, my first exposure to the particular Berkeley culture, and exciting it was, too. The dark corridors of Wellman Hall housed faculty, students, the Entomology office staff, the Essig collection of insects, a library, and a teaching lab, and it was in Wellman that I attended meetings of the department at intervals and went to the chair of the department to complain.

My lab, however, was three miles away, at a ten-acre site called the Gill Tract, because it was here that the Division of Biological Control had its headquarters, and as a researcher on plant-insect interactions, I was to work on the biological control of weeds. At my interview, I had been somewhat surprised at the run-down nature of the place, but in my excitement I looked on the bright side, imagined change, and was sure things would be fixed up for me if I were to get the job. In any case, there were plenty of old rooms full of junk that could be cleared away.

When I arrived, however, I was given three small rooms. They were totally empty. I was so surprised that it took me a few weeks to realize this was it. I was informed that I would get ten thousand dollars in Hatch money, but couldn't expect anything else. I realized that there was no point in getting mad and that I should have negotiated. This culture was brand-new to me; I was a novice. No longer did an appointment automatically mean the provision of the wherewithal to do the work.

I used much of the limited funds to equip my lab from all the second-hand furniture and equipment places I could discover. I bought a fine wooden chair made by prisoners at San Quentin prison (as I learned from a plaque underneath the seat) and I cadged microscopes and lights from a professor nearing retirement. I was left completely to myself, and, in the absence of any project or money, I borrowed a video camera and tape recorder and embarked on something I had wanted to know: can grasshoppers learn and, thereby, improve their foraging efficiency?

I did realize that my naïveté had been a problem, but I decided I would show them and use all my ingenuity to demonstrate I could operate, and do interesting research, even with almost no resources. And I would follow the topics that interested me before I embarked on biological control, especially after a more experienced expatriate in the department said to me, "You have tenure, you are a full professor, you can do anything you like. Just do something interesting and become famous and you will please higher levels at UC. That's all that matters!"

Others said, "You need to write grant proposals to get equipment," so I submitted three in the first month, after having several of my new colleagues read them. "Great," they said. "Wonderful, excellent." And so, I had some hope. Later, I discovered the proposals were ineffective by the accepted standards and that no one had honestly advised me, which was perhaps the hardest lesson for me in that first year. I was to make a fool of myself with the National Science Foundation and the United States Department of Agriculture review panels, even with the administrative staff in Berkeley.

In time, I did learn the tricks and obtained grant money, but not in those first years, and it wasn't easy. First, I managed to get money from a private foundation, on the advice of a good old friend on the East Coast. Then I discovered one could get funds to improve unsafe working conditions, and so I started work on plant chemistry that required the use of toxic solvents. For this, I took my extraction apparatus out on a cart into the yard at the Gill Tract so that I could write a report on the extreme situation of the conditions I had to work in. This got me money to roof in a space between two buildings and make a new laboratory (without heating), where a fume hood was constructed to satisfy health and safety regulations. I was learning to be more resourceful.

But my lab remained minimal. I politely complained, but the Division of Biological Control had no money to distribute. I went up to the main campus, past the pollarded trees and into Wellman Hall, to see the chair of Entomology. His response was, "You are in Berkeley now, what more do you want?" I could have been angry, but at some level I enjoyed the challenge and decided I would be damned if I couldn't get what I needed. Meanwhile, I continued research that could be done with minimal equipment, and I hardened my resolve to work on whatever pleased me, never mind what my chairman might expect. I was learning to be an egotistical American and, at the same time, I was enjoying life in the Bay Area of California.

One day I had lunch with an elderly faculty member. He did no research himself, but he liked to pontificate. He said to me, "You don't want to collaborate with anyone. You need to make your own mark with single-author papers." I was aghast. It had always seemed that teamwork was the best way to do research. My experience as a government scientist in Britain involved much cooperative activity, but there had not been the fierce competition for money that was to be part of my career in the New World. However, I understood his message and kept it in mind as I evolved my own research effort and worked around the problems— greenhouses with missing glass, environmental rooms and cabinets from a previous age that were all out of order, screen houses with torn screens, insect-rearing rooms in disorder, inadequate quarantine facilities, leak-ing roofs, broken-down trucks, rooms full of unlabeled insect materials from the good old days of biological control greats—Drs. Messenger, Van den Bosch, and Huffaker. I had worked in several developing countries, and the Gill Tract was just like them.

Within the first few months, I discovered that there was a kind of biocontrol religion. The central tenets included the following: too much theoretical work was counterproductive; biocontrol and inter-cropping were the only ways to deal with insect pests; chemicals of any kind were wicked; genetics was a load of baloney; molecular biology was evil and took all the money; Latin America was a good place; the other biologists, including entomologists on campus, were out to deprive the biological control faculty of money and facilities.

As a nonreligious person, I found myself needing to go to the main campus more often. I wanted to enjoy all the seminars and journal clubs. I was exhilarated by the often-wild ideas, the wide-ranging discussions, and the novelty of getting outside the box that had been my training in England. There I was thrilled to find an environment so suited to the person I felt I was, where imagination mattered more than discipline. I bought a bike and cycled to and from the Gill Tract, chaining my bike under the pollarded trees, to find collaborators in the wider campus community. After my years as a government scientist in Britain, I was discovering the joys of academic life, freedom, and American individualism.

One day, I discovered that a newly hired young assistant professor would be given over one hundred thousand dollars in startup funds, and I decided the time had come to make an appointment with the provost to complain once more and request a dollar amount to complete my modest setup. She was sympathetic and immediately made available all the money that I asked for. It was a momentous occasion, for I had learned to fight. I had learned to push for myself and to get what I wanted instead of politely waiting for anyone to help me.

I went to see the chairman of my department—after all, I would be bringing money in and he would surely be pleased. Across the lawn I went, past the pollards, up the steps of Wellman Hall two at a time, and into his office. When I told him my news, he looked at me speechless. Eventually he said, "Because you are a woman, I suppose!" I was too taken aback to reply, but I never forgot that jealous response. So much for liberal Berkeley, I thought.

Afterwards, sitting in the Campanile Esplanade among rows of pollarded plane trees and pondering the novelty of being a professor at Berkeley, I realized that the symmetrical planting seemed appropriate for the neoclassical architecture on parts of the campus. As I was thinking about the paradox of that liberal institution—with faculty who could remember the free-speech movement of the sixties yet had limited ideals of equality—I was also examining the details of its lovely campus. Later, I read that the classical ideals expressed in designs by the Paris École des Beaux-Arts began to appear on campus with John Galen Howard's plan. He designed Wellman Hall and other buildings and organized plantings with architectural patterns in allées, bosques, and hedges, and, with them, the pollarded plane trees. Funny, I thought, how the trimmed trees represented a kind of tough discipline that had been my student training, while now I was learning to branch out, though I still had much to learn.

The learning curve had already been steep. I thought back to my interview and how incredibly naïve I had been about anything to do with the university system, the Bay Area, and the United States. I'd had a meeting with the dean of the College of Natural Resources. My ideas of college were based on either what I had known at the University of London, where each college is effectively a small, semi-independent university, or the system in the University of Oxford, where the colleges are independent of the academic departments. What was the College of Natural Resources, I wondered, apart from the obvious fact that it included the Department of Entomology?

I kept my ignorance to myself. At one point the dean remarked, "We are a land-grant university."

"How wonderful," I replied, conjuring up the rich Californians that must give so much land to their famous institution.

"And we have the experiment station," he continued.

I was thrilled with this, too. There must be some big farm or something for doing field experiments. Later, when I realized that the experiment station was not a place at all but an administrative structure relating to a federal government act in the late nineteenth century, I was somewhat dashed. But that was long afterwards, and it no longer seemed to matter. It wasn't for several years that I properly learned about the Hatch and Morrill Acts, which allowed Congress to make regular appropriations for support of agricultural study and land grants to form agricultural colleges within existing universities or to build new ones. Experiment station meant Hatch money administered separately from the university budget and research related to agriculture.

I thought back to my days in England and in the African and Asian countries where entomological work had taken me. And I thought back to my earlier college days in Queensland. Everything had been fun and interesting, and, culturally, it was easy to slip from one country to another. I was a British citizen and, before that, the citizen of a country that began as a colony of Britain. Coming to the United States was something altogether different. The transition had happened almost by chance, because I knew little about the big country across the Atlantic, and my European colleagues were mostly critical of it. "They kill one another with guns," some said, or, "Think of the KKK, for goodness sake." I had applied for the job in a moment of frustration with government red tape, was surprised when I was asked to come for an interview, and was quite taken aback when the professorship was offered. But, adventurous by nature, I took the plunge without even thinking of what it would be like. And Reg was with me—a devoted husband who gave up his career in order for me to develop mine.

I was learning about individualism, a trait that is well developed in Americans and amply found in Berkeley. It has to do with that all-important independence, confidence, and much-touted freedom, but also with the need to look after number one first. At least, that is how it appeared to me. Confidence abounded among students and faculty, politicians and secretaries, shop assistants and street people. Richard Rodriguez writes that the American has "the confidence of an atomic bomb informing every gesture," and I found this confidence fun and catching, even if it was sometimes confused with intelligence and even if it was often unwarranted. No longer necessary was the rigor of strict rules or the requirement for anyone to give up anything to fit overall expectations, or the community, or the institution. Discipline of behavior and criticism of people and values were uncommon. Unusual logic, beliefs, and behavior were all accepted. There seemed to be a desire to include everything in the worldview. Terrorism had not yet struck.

No classified social system existed as far as I could tell, other than that of money, although there was an element of superiority about being associated with the University of California, Berkeley. Joan Didion in *Where I Was From* writes that at age thirteen, she asked her mother to what class they belonged. Her mother replied, "It's not a word we use. It's not the way we think." It was this, perhaps, that seemed most refreshing to me coming from Europe, even though the class system in England was a lot subtler than Americans imagined. My New World friends like to point out that California may be different from other places in the United States, but Philip Roth had it right, I believe, when he summarized the difference between liberal America and more exacting Europe as "everything goes and nothing matters" versus "nothing goes and everything matters."

Now, more than thirty years after arriving in Berkeley, most of those years spent in Tucson, Arizona, I look back with fondness to the

California beginning, the home of my first six years in a new country. How I learned to be independent, more egotistical and demanding, yet accepting of others' ideas and lifestyles. Instead of the rigidity of rules that had generally made for smooth social interactions in English life, flexibility became the order of the day. Each individual had a say, and a story: a student could take an exam early if travel plans would be interrupted by waiting for the big day; a shopper at the store checkout had the right to discuss all manner of things with the cashier, even if there were a queue behind him. There were pros and cons, but I enjoyed the new life.

I became broader in my thinking, more flexible in my intellectual pursuits. I had to learn the rigors of writing a convincing research grant and explaining why the work was important, a process that involved the self-advertisement that had always been anathema to me, yet ended up making me feel that I had a distinct place in the scientific world. The seminars and discussions, colleagues and students took me into unknown places, showed me new ways to see theoretical problems, and ultimately made my work more rewarding and more visible to colleagues round the world. I began to feel the confidence that had seemed so unrelated to my life before I came to the United States.

Perhaps because of this new freedom and confidence, I developed many new research projects and learned the joy of working with graduate students to think about new ideas and solve problems related to the lives of insects that feed on plants. I began to design my experiments more clearly within the bigger pictures of ecology and evolution.

I loved the green campus with its classical architecture, its groves of trees, the view across the bay to the Golden Gate Bridge enshrouded in fog. Those pollarded trees in the pale light of the western sun, the misty mornings that couldn't be more different from the Sonoran Desert where I live now, or the landscape of southern France, those willows painted by Van Gogh. The luminous picture in my kitchen, the brilliant colors, the excitement painted there on a few old pollards, is a reason for standing still a while as I pass, for letting my eyes rest on a scene that at once is France and a particular painter, but also, for me, the University of California, Berkeley campus and all I learned there about becoming an American.

A Taste for Novelty

The midsummer sun bleaches the colors in the southern Arizona desert, and my eyes feel weak. It is noon, 105°F in the shade. A grasshopper crawls up the trunk of a mesquite tree and settles to rest among its foliage. It is a fine specimen, as large as a mouse and glossy black. Yellow marks trace the outline of a jacket on its thorax and on the helmet of its head, while its front wings sport bright green veins. Its accepted common name is the horse lubber, but it is sometimes called the Mexican general. The one I watch is a female, as I can tell from her large size and the four points of her egg-laying apparatus forming the ovipositor at the tip of her abdomen.

I have been watching her every move since 7:00 a.m., when she started her day. Plants rejected or accepted, time spent in each feeding bout, movements between plants—all details I have fed into my handheld computer. She has been on the move almost nonstop and has fed on more than fifty different plant types, challenging my attentiveness and endurance. As temperatures rose to 100°F, her movements became faster, her feeding furious, but at last, it was too hot. Needing to escape the fierce ground, she made for the breeze above in the mesquite tree. In a couple of hours, she will descend for her long afternoon of foraging, and I will continue my recording. I can predict this because for eleven days I have been standing out in the summer sun, watching individual lubbers for ten hours at a time.

You wonder why I do this. In the heat of the afternoon, I sometimes wonder, too. But I burn with curiosity when I try to answer questions about the behavior of animals in nature. I know that individual lubbers eat a wide variety of plant species. It is an unusual strategy for plant-feeding insects; moving about, they expose themselves to discovery by predators— birds and lizards. My questions are twofold: Why do they march about eating so many different things? And by what process do they make a decision to switch to a new food? So I study behavior in detail for clues. I watch and record and think. For ten hours each day, I almost become a grasshopper.

Three months later, I have some answers. I have analyzed my field observations and I have performed laboratory experiments. These big, shiny black creatures have a lifestyle as distinctive as their appearance. They have a preference for poisonous plants such as *Euphorbia*, upon which they have long bouts of feeding, while they disdain the blander grasses and wild lettuce. In captivity, if they have only one plant species available, they sometimes become ill, evidenced by lethargy or even collapse. Eating many different plants, it turns out, prevents the level of any single plant chemical from reaching a poisonous dose, just as with people; for example, there was the man who died from eating nothing but carrots, because too much vitamin A is toxic.

In addition to diluting potential toxins by wide-ranging foraging, horse lubber grasshoppers absorb poisons out of their favorite plants during digestion, then accumulate and store them in special glands. If attacked, they lift their black and green front wings, flash their red back wings, buzz, and exude an ill-smelling, frothy, poisonous mixture. Predators who make the mistake of going for these grasshoppers spit them out, and, remembering their unique sounds, smells, and sight, ignore them in the future. So, I have an answer to question one: broad eating habits enable the grasshopper to dilute toxins as well as sequester a mixture of really nasty ones.

Interesting as this may be, the second question is even more exciting—what is the mechanism for switching between different foods? These grasshoppers are surprisingly good at learning; they can be taught to find food in boxes in the laboratory. Yet while foraging in nature, they eat too many different plants to be able to associate the quality of any one plant with its value or toxicity, so they cannot learn to choose well. However, if given a synthetic food with a simple flavor—say, vanilla—they will eat it for a few short bouts, but then reject it. They will accept the same food if the flavor is changed to, say, peppermint, but again, only for a few bouts. And after a while on peppermint, they again accept vanilla. In short, they can be tricked into eating and growing well on nutritionally identical foods, provided that new flavors are added at intervals. Their decisions to change food are based on the desire for a new flavor. Just as people do, they like variety!

Because I am a biologist, I write up the work, carry out statistical tests to quantify the results and estimate probabilities, then send it for publication in a scientific journal. One anonymous reviewer slams it, but another deems it acceptable; the editor decides to publish it with minor revision. You can read it in the journal *Animal Behavior*. In exploring an interesting question, I found an answer that adds, in its minor way, to biological knowledge.

For many years, I have been studying insects of numerous kinds. Sometimes I've investigated a question about how to control a pest or simply about its basic physiology or ecology. Sometimes I address a question because it's just plain fascinating, with the endeavor having no obvious benefits for humankind. I have worked in Australia, Nigeria, Mali, India, and England as well as the United States, and I've taught students from around the world, all wonderfully different. In short, I have enjoyed immense variety at all levels. I am one of those for whom variety really is the spice of life, so I find particular pleasure in my study of the lubber grasshopper. I feel an affinity with my lubber, not only because I love its flashy appearance or fancy chemistry, but for our shared love of novelty.

Unlike me, my colleague and beloved partner Reg liked consistency. For example, he had twenty pairs of socks, all the same color. If I made stew that lasted a week, he wanted nothing else, but I prefer to have diverse foods. Each morning when I get up, I like to choose between yogurt, fruit, egg, toast, or any other thing for breakfast. I love to choose. By contrast, he had cornflakes every morning, always. If we discovered a nice wine, I looked to buy a bottle, he wanted a case.

I became obsessed with the contrast and started noticing the habits of other people. There is a big spectrum, from those like me who love variety in everything to those who are most comfortable with complete predictability. So what, you ask? Well, I can't help wondering if there is any biological relevance in the variation. You need to think less if you are consistent, therefore the development of regular habits leaves the mind free for other activities. If you like to have choices, then you must spend time choosing, sometimes long periods of dithering. But then again, does it matter?

Back to grasshoppers. The horse lubber has little to worry about with respect to predators. It can take its time, make choices. By contrast, the creosote bush grasshopper, *Bootettix*, is so well camouflaged, it is very hard to find, and it feeds only on creosote leaves. It has no choice of foods and moves little, but it is potentially very vulnerable to predation, because it is a species not chemically protected. Many plant-feeding insects lie between the two extremes of lubbers and the creosote specialist, feeding on various plants, but with no chemical protection. Here, perhaps, the tradeoffs in behavioral patterns become plainer. On the one hand, they could feed on only one plant type and retain vigilance against predators; on the other hand, they could enjoy choosing the best food among plants of differing quality, with the possible cost of not being able to pay full attention to danger. In other words, the choice is between safety versus optimal food. These kinds of tradeoffs lead to the maintenance of varied behavioral patterns, because each strategy has a benefit, and the benefits vary depending on the circumstances.

We can all think of tradeoffs in the strategies of people. For example, one can be jack-of-all-trades and master of none, or one can be an expert in a few tasks. Each strategy has its value, depending on the situation, and they can be mutually exclusive. Perhaps the commonest behavioral tradeoff we are aware of in people is between doing a job slowly, carefully, and well, but making slow progress, versus doing the job fast and a bit sloppily, but getting a lot done. I might want the first approach in a nurse and the second in a house cleaner.

Perhaps it's difficult to assign biological significance to the variation among people with respect to the love of variety, but I imagine the value of different degrees of novelty seeking was plainer in our evolutionary past. Did the gatherers keep an eye open for new foods and have a yen for novelty, while the hunters of animals kept strict attention to well-known paths of thought and action? Were the individuals with an eye for novelty the ones who discovered new and potentially useful tools and plants, while those with well-developed habits were the ones most skilled at familiar jobs? Today, perhaps the inventors of computer programs are the novelty seekers, and the consistent users of those programs the best at getting down to work. Psychologists tell us that novelty seekers are more likely to engage in dangerous activity or take drugs. They don't tell us what might be the tradeoff, or potential value, though some sociologists claim there is a relationship between some risk-taking behaviors and attractiveness to women. For example, the sight of men involved in dangerous sports is said to excite many women and girls, which may, in turn, provide a benefit for the men. For my part, I think the risk takers probably have more fun, attract or meet more potential mates, and are more likely to chance upon the bonanza!

In discussion of generalists versus specialists, biologists can always find reasons why specialists should evolve—they can simply be better at doing whatever they do and outcompete the less skilled. Explaining the evolution of generalists, or understanding what benefits allow generalists to exist in nature, seems to me to involve the act of trying things out, taking risks, giving it a go. An unrecognized, but wonderful, food or mate or opportunity may be the reward. Thus, a successful generalist may be the ultimate novelty seeker.

The parallels between people and other animals, even insects, are remarkably strong. Twenty years ago, animal behaviorists resisted thoughts along these lines and were critical of any hint of anthropomorphism. And I know that an objective approach is necessary for clear-cut experiments repeatable by others. But the atmosphere has changed. It is not that we can afford to be less careful, but that the genetic revolution has demonstrated the common ancestry of all animals. Genes have changed little over millions of years, though new species carrying them have evolved. The embryo of a fly with faulty eye-development genes can make proper eyes if mouse eye-development genes are inserted; genes for behavioral traits can also be similar among animals.

Genes can now be viewed on the Internet, and each day, from laboratories undertaking widely different research around the world, more of them are posted. If you or I determined the structure of a certain behavioral gene, we could look it up and see where else it has been found and what has been discovered about its function. In a matter of minutes, we could see if the gene involved in novelty-induced locomotion in rats is the same or similar to a gene for risk taking in humans, to one for rover-versus-sitter behaviors in fruit flies, or to the gene for shy-versus-bold behavior in sunfish.

My feeling of closeness to lubber grasshoppers is not, perhaps, crazy. Indeed, the fact that I share behavioral characteristics with individuals of other species, and such different ones at that, gives me the profoundest sense of belonging in the world.

The Hawk Moth's Progeny

Her name is *Manduca sexta*. She is a big, gray moth flying fast like a hummingbird. She is attracted to the sweet smells of night-blooming flowers. She lives for a week or so after emerging from the pupal stage in a cell in the ground, a cell the original caterpillar made at the end of her feeding life.

My initial observations of her occurred by chance in 2001, after dinner on the ramada outside my bedroom overlooking the desert. The first night, she flew to the great, white trumpet flower of a nearby sacred datura and hovered there, unrolling her three-inch-long proboscis and drinking nectar. Then she hovered elsewhere round the plant and curled her abdomen many times, on each occasion depositing a pale green spherical egg on the undersurface of a leaf. From then on, I spent many hours watching her behavior. She rested nearby during the day and, on successive nights, laid 210 eggs in all. A screech owl took her in the finish.

A small parasitic fly that laid eggs of its own inside moth eggs found twenty-seven of the *Manduca* eggs, and, about a week later, with a magnifying glass I could see four tiny flies ready to come out of each egg. Earwigs crawling about on the plant at night ate a few eggs. I saw one doing it and noticed the characteristic mess of jagged eggshell left behind. Delicate stilt bugs that normally suck plant sap, but apparently need an egg or two of some hapless moth in order to develop their own offspring, sucked out seven eggs, leaving tiny holes in the shells. Lacewing larvae that ingest all kinds of small insect material, inserting their pointed jaws, each with a suction tube down the middle, ate four eggs. Ants carried off about fifty eggs, all sorts of them patrolling the plant during the day and night. There were perhaps a few other predators that went unnoticed, and a few eggs just didn't hatch. They may have been pierced or diseased or may simply have been unfertilized.

Exactly 113 eggs hatched into little caterpillars. They ate their way out of the shells, leaving half shells that lasted for weeks. The caterpillars began feeding on leaf tissue within a centimeter of the empty shells, each one making a small hole with its first meal. One day later, there were sixty-three left. Ants, lacewings, and soldier bugs took most and a few scavengers took some half-eaten remains.

In the next three days, as the remaining caterpillars ate bigger holes and moved about more, the numbers were reduced to thirty-one by different kinds of sucking bugs, various ant species, lacewings, crab spiders and jumping spiders, and even male web spinners, which are normally thought to eat plant detritus. Twenty-four grew and success-fully molted to the next stage, with heads twice as big as the first-stage caterpillars. But the depredation continued. Spiders and ants took most of the second-stage caterpillars, with hunting wasps, the kind that fly about and take caterpillars away to feed their own young, collecting two. Seven caterpillars made it to the third stage. I saw a cactus wren eat one, and a big reduviid bug tackled another with its sharp proboscis, causing it to fall to the ground mortally damaged. Ants and spiders went on with their foraging. Only two caterpillars molted to fourth-stage larvae, one of which was taken by a thrasher, so that a solitary caterpillar became a fifth-stage larva, growing to be three inches long. It ate nearly all the leaves of the sacred datura plant. I decided to keep it and rear out the adult.

I let the big larva evacuate all its gut contents, which it does at the end of larval life, and then I put it in a box of sawdust, where it buried itself and turned into a pupa, preparing to undergo transformation to a moth. I kept it warm by placing it beside the hot water heater in my house. Months later, thinking the moth should have emerged already, I emptied out the sawdust. The pupa was black and shriveled. I found an additional two small pupae that turned into flies. My caterpillar must have been attacked by these parasites before it finished feeding, the maggots having eaten it out and killed it. The moth I'd noticed months before had no surviving offspring—zero fecundity, though she was such a fine specimen that laid over two hundred eggs.

One day I told my friend Janet about my *Manduca* observations. "My god, Liz," she said, "isn't this a dilettante kind of occupation? Who cares about a moth?"

"What is more," I laughed, "it takes hundreds of hours to find out and document the stuff."

Janet looked at me in silence for a minute or so, and I knew she was amazed, and from the perspective of her vocation as a nurse, I could understand that. She fired a series of questions: Why did you do this? And do you really mean to go on doing it for the whole of two Septembers in southern Arizona? And are you really going to inveigle a nice visiting Spaniard to come and do it with you? Why does he think it is interesting enough to spend hours watching?

I tried to answer, "Well, it's fun. I want to know what happens." But even as I spoke, I knew I was questioning myself about why I was quite so interested.

Why is one curious about anything at all? Why does one develop an interest in searching anything in depth? For me, it goes back to my childhood, when I worked in the garden with Mama. She encouraged me to look at details, to press flowers, catch ladybugs, and germinate seeds. Looking makes one wonder, and I think that is where the curiosity began. One memory is particularly sharp: Mama and I knelt together by the side of the long flowerbed, digging into the rich earth with our garden trowels, turning and crumbling it. We made rows of holes, and I moistened each with a dribble of water from the hose. She unwrapped the parcels—first a roll of newspaper and then a piece of old hessian—exposing seedling plants with their roots embedded in damp sawdust. She lifted one free with her left hand and suspended it in one of our holes, and, with her right hand, she carefully pushed earth around the roots until the plant stood firm. She patted the earth down all round it and added more water. We worked together and knew our love.

"Mama, how *do* these flowers grow?"

"Just from water and dirt and air."

The thought captured my eight-year-old imagination completely. I already loved the little pink and cream flowers of linaria, the dark blues of lobelia, purple delphinium, orange wallflower, but after this revelation, each plant was special. When Mama wasn't looking, I stroked leaves and hugged the jacaranda tree. I knew that living plants were amazing.

Then there was grade school—the strange feelings of elation when suddenly there was the picture in my mind of why a light comes on when the switch is flipped. I can even remember the bursting sensation of my beating heart and the warmth of my flushed face that was not due only to a hot Queensland summer day. Imagine that two gases, oxygen and hydrogen, when properly combined in the ratio of one to two, make water. And how curious and wonderful that even solid objects are made up of particles with infinitely more space between them than the size of the solid matter itself. The fact that one could find out how complex things worked was a revelation, however oversimplified it may have been at the time.

Many people find it most peculiar that anyone can watch animals continuously for hours, recording behavior and watching the minutia of the world, enduring long periods of discomfort in the process. But I do enjoy nature, and I am elated when I think I may perhaps answer one of the larger controversial questions that fascinate biologists in general. Perhaps if one takes a novel approach and puts in the time in just the right way, one might turn tides of opinion, take ideas to a different level, upturn conventional wisdom. There's a thing—prove the experts to be wrong!

Controversy enticed me into the study of *Manduca* after chancing upon that moth laying eggs on my datura. Biologists had been arguing about the relative importance of the food plant versus predators and parasites in determining the affiliation of a plant-eating insect with its particular plant host(s). Plant-eating insects and their host plants make up 50 percent of all species of life, if one excludes bacteria and viruses, and most of these insect species specialize on just one or a few types of plants. Why?

I turned over the ideas in my mind for years. Having the time for reflection during observation is another element in the joy of observational research, stewing together curiosity, known facts, conventions and hypotheses, possibilities and doubts, until new connections arise and eureka moments emerge. I became consumed with the thought that the current convention of studying only the host plants and their insects was unsatisfactory; natural enemies of the insect herbivores must also matter.

For fifty years or so, the predominant opinion had been that plant and insect evolutionary histories are tightly knitted together in a kind of arms race. A plant species evolves a new toxic chemical against its insect enemy and, when so protected, expands its geographical range and evolves into a variety of new species in different habitats. In time, the relevant insect species evolve ways to circumvent the defenses, reoccupy the plants, and themselves evolve new species on the new plant types. This circular process was called chemical coevolution. It seemed to explain specialized feeding behavior and also to provide an engine generating diversity.

I accepted these popular beliefs for some years, until I noticed details that began to alter my view. For example, many so-called plant-defensive chemicals deter feeding but are not toxic to the insects if they are force-fed to them. Therefore, an insect not liking a plant was not linked necessarily to its poisonous effects—its so-called defenses. An insect's dislike of a plant may evolve for other reasons, such as it being more vulnerable to predators on that plant. In Berkeley, I was able to show experimentally that specialist caterpillars preferring just one food-plant type are less likely to be taken by predators than are generalist caterpillars that eat lots of different plants. This gave me the cue to study the interaction of plants, herbivores, and their natural enemies in concert—what's known as tritrophic interactions. And I couldn't resist the opportunity that chance provided in the form of *Manduca sexta*.

The hawk moth I had noticed in my garden had produced offspring with 100 percent mortality from natural enemies before reaching maturity.

How common was that? What did that mean for the current theories of plant and insect interactions?

A devil's claw plant was growing close to the original Datura plant, and the next moth to visit laid eggs on both plants. This was odd because devil's claw is in a plant family, Martyniaceae, not known to host *Manduca*. My hero moth is a specialist on plants in the potato family, Solanaceae, including tomato, datura, and tobacco. And here is where the story broadened. I compared the fate of eggs on devil's claw to those on datura, and found the same range of predators, but while survivorship on datura averaged only 0.4 percent, it was all of 2.0 percent on devil's claw, making the latter plant five times better for survival. No doubt about it, natural enemies were very important. Such a major difference matters enormously in evolution. Would *Manduca* change hosts and become a devil's claw specialist? If what I found was typical, then it should, because individuals favoring devil's claw would produce so many more adults that devil's claw–preferring moths would quickly predominate in the population.

To extend the study, I needed a work companion, and I found Spanish Alex, who was working in the lab of a colleague but who was also fascinated by my story. We talked about Spain where, in my past, I had many wonderful holidays. And we talked of plants and insects and their evolution. Wiry and dark with excited black eyes, he jumped at the idea of the *Manduca* study and a chance to see more of Arizona. Together we planned the study. We paired up specimens of datura and devil's claw in Catalina State Park, the Santa Rita Mountains, and the Patagonia Mountains, all close to Tucson. We spent many hours traveling, searching plants, looking at eggs and larvae, doing a Sherlock Holmes on what had taken place between visits.

"Look," Alex would cry as he peered through his magnifying glass at the underside of datura leaves. "One egg pierced and sucked out, six eggs messily broken by a chewing predator, a group of eggs with developing parasites."

"And here," I would say, "three leaves almost all eaten and a big stain from a vomiting caterpillar tackled by some big natural enemy." The caterpillar was gone of course, but the evidence was strong. Then Alex would bound over to the next plant with the energy and excitement of a ten-year-old.

"Hey," I would say, "what about this! Twenty-seven ants and lacewing larvae and parasites embalmed in sticky goo on devil's claw."

"Yeah, and look at this miserable little larva—just managed to make a hole of about two square millimeters and kicked the bucket."

We found that there was a pattern. Devil's claw is not a great place to live, but it's better than datura, and, sure enough, the causes of caterpillar mortality were different on the two plant species. The predator pattern on datura was similar to my initial observations in all cases, but on devil's claw, the picture was entirely different. The sticky leaves of devil's claw entangled all but the most dexterous predators, while *Manduca* larvae ate through the sticky secretions with apparent abandon. On the other hand, a noxious gas weakened developing embryos and baby larvae on devil's claw; we found dead eggs, larvae too weak to get properly out of the egg, or thin little hatched larvae scarcely strong enough to start feeding. The effect was profound in hot, dry conditions, yet minimal in rainy weather.

Numerically, it turned out that for maximum fecundity, moths should lay eggs on datura during the hot, dry blasts, but on devil's claw when moisture levels rose. In general, overall survivorship on the novel host, devil's claw, was significantly better than on the ancestral host, datura. Why then did this moth continue to make use of datura? It was not my original question, but science is like that—things take an unexpected turn and one must follow where the new path leads. In fact, most of the interesting findings in biology start with something unexpected, something out of the ordinary, where someone or other decided a typical thing was worth another look in a new light and ended up seeing the exceptional rather than the expected. That is how Darwin got onto evolution.

The plot thickened as we took on a second year of work. *Manduca* moths emerge in spring to mate and lay eggs. The larvae complete development, and a new generation of moths emerge in summer. These in turn lay eggs and the pupae that finally make it stay underground for the winter, to emerge as adults the following spring. In other words, there are two generations of moths each year. Here is the problem: the ancestral host plant, datura, is a plant that is available spring and summer for the moths, whereas devil's claw is an annual that comes up only after the summer rain in August. Clearly, *Manduca* cannot switch to devil's claw as the sole host, because there would be no plants available for the spring generation. Female moths need to stay with datura, but in the wet summer months, they must include devil's claw in their repertoire of hosts to enhance the numbers of surviving offspring, especially in rainy weather. And this is what they do.

The story of interrelationships in ecology, the intertwining roles of herbivore, plant defenses, natural enemies, and weather, and this microcosm, excited me at intervals for two years. Alex, now back in Spain, emails me from time to time about the lovely *Manduca* days in southern Arizona and the tale of tradeoffs that ecologists love to discover and to talk about. For me, our finding is a good example of the tritrophic interaction between predators, caterpillars, and plants. It is also a great lesson in how an initial observation can become complex with study, similar to Darwin's tangled bank. But above all for me was the thrill of finding out how something works—an excitement that is with me still.

Jaws

I said to my graduating PhD student Dave, "How would you explain your entomological research to someone who just happened to be sitting next to you on the bus and wondered what the hell you were doing with public money?" He was taken aback and finally said, "Give me a few minutes." In fact, he never did come up with a good answer, and I decided to ask the question in every PhD oral exam thereafter. Some students explained potential economic reasons for their research, like controlling pests, while others said the training was of general importance to their development or for getting a job in biology, but I really enjoyed the answers that focused on how engaging they found their study. I suppose that is because I am enthralled with every research project with which I become involved.

I am a generalist and let ideas flow through my head making their own paths, stimulating unknown or unconscious corners of my brain, interacting with thoughts from totally unrelated events. My approach to science is more on the creative side than on that of logical deduction, mathematical reasoning, or strict hypothesis testing. The invention of possibilities to be proved or disproved is something of an obsession. When I do have a hypothesis to test and results prove something quite different from my expectations, I am thrilled with where the idea might go next, what new window might open, what novel connection might be made, what innovative idea might emerge from it.

As I sit on my porch in Tucson watching the wind sway ocotillo branches, I wonder about the strength of fibers in their stems. Such a strange plant, too, with long stems all arising from the base, and a habit of bearing leaves briefly after rain. One rarely sees broken stems unless they are dead. Instead, they sway and bend, sometimes to angles of ninety degrees. They commonly don't have leaves on them, and I wonder if leaves influence the bending angle. I think of those rows of silk-soft little emerald leaves and realize I have never seen any sign of insect damage on them. Is it because they are so ephemeral that there is never a temporal match with particular insect species? Maybe they are filled with strange chemicals, bitter to herbivores. A cactus wren lands on one of the branches, bending it down by ten degrees, and as his angry-sounding argument fills the air, I see something small land on the branch just below him. It is half an inch or so long, but I will never know what it is because the wren pounces, and it is gone. I note that these long, thin stems, without complex architecture, hardly allow for insect life to hide. With a good population of insectivorous birds and small herbivorous creatures nearby, they don't have a chance.

I remember the walk I took at Catalina State Park one weekend, all those spring-fresh blades of grass, though most of it is troublesome wild oat introduced from Europe. I pulled a stem of *Aristida* grass to put in my mouth and reflected on its texture: tough, hard, chewy, fibrous, and, embedded in the cells, a multitude of tiny silica particles, minute sand grains. No surprise that grazing mammals had to evolve teeth with enormous grinding surfaces. Apart from having stems to chew, grasses—or, to be precise, grasslands—seem to mean something special to people. Parks and sports fields; swaths of green on mountainsides; great golden savannahs in California, Kenya, Australia; prairies in the Americas—grasslands are so commonplace and so much a vital part of our scenery. Is it our evolutionary past on the African plains that somehow affects our emotions?

Lawns, too—afternoon tea on the back lawn in Brisbane, Queensland, with Mama pouring from a silver pot into her favorite Japanese cups with saucers. There was a sponge cake she had made in the morning, and the family lounged around on Great Gran's patchwork quilt. Brothers practiced cricket on the lawn, bashing into the turf with their bats. My father mowed it, adding to the sounds of hand-pushed lawnmowers clacking through Saturday afternoons. My sister Jennifer played lawn tennis. Puppies rolled and played on the front lawn, while I bicy-cled with other neighborhood kids down the sidewalks that were completely grass covered. Here in the Arizona desert, there are many golf courses. I spat out my feathered stem, thinking of teeth, of grass blades, and of the grass-covered beauty of the English Lake District, where Reg and I had walked while chewing the sweet stems of fescue grasses.

Leaf texture has interested me for many years. Gardening with Mama when I was seven or eight involved detailed inspection of plants, and, even then, I was surprised to find so many different sorts. Look at leaves in detail, put them under a magnifying glass, and one will find that each species has its peculiarities, its special pattern of veins and gaps between veins, of shape and thickness, hairs and spines, knobs and wrinkles, glands and pores, blooms and threads of wax. Within, one finds a diversity of crystals, fibers, silica particles, mucus, miniature sacs of toxins, idioblasts, latex vessels, starch grains, raphides, sclereids, cystoliths. The arrangement of cells that do the work of photosynthesis and transport varies, the amounts of lignin toughening the fibers vary, the shapes of all the tissues vary. Everything varies.

I think of my student Frank, who studied caterpillars of the California oak moth, *Phryganidia*. Frank, the perfectionist, who needed to travel every path to its end, or at least to the point where it could no longer be seen. Frank, who never saw the unexpected, who did meticulous work, and whose research tuned in to the minutia of leaf anatomy and to the very fine details of how caterpillars manage to eat the tough old leaves of California live oak.

The mother moth lays a clutch of eggs on the underside of a mature oak leaf, where they are protected from the force of rain and the heat of the sun. Tiny caterpillars with relatively enormous heads emerge after about a week. The little commas crawl to the upper surface of the leaf to eat and return to the sheltered undersurface between meals. They have problems. Between the tough networks of veins on the surface where they feed are tiny patches of smoother leaf surface from where they must extract the soft green parenchyma tissue within, but the size of these patches is smaller than the diameter of their heads. To get there, they must also chew veins, and it is for this reason that their heads are so large. Most of the head cavity is filled with the muscles used for closing their jaws; bigger heads house bigger muscles and provide the strength required to sever veins of California live oak.

Frank was not given to seeing the big picture. Rather, he delved deeper into the particulars. He was always very polite and formal, the only student I ever had who could never call me Liz. "Dr. Bernays, so sorry to bother you, I would like to investigate the details of veins within the leaves. Please would you show me the techniques for embedding and sectioning them?"

"Of course," I replied. "Let's do it now, but remember to look at that other little caterpillar, *Bucculatrix*. I think it has a different strategy, and the comparison might tell you something. Oh, and have you designed experiments to determine why the little *Phryganidia* larvae must feed on the upper surface of the leaf? What makes them leave the shelter of the undersurface? And what makes them go back, for that matter?"

Frank would give me his shy smile yet again and look down, below enormously long eyelashes. "Oh, Dr. Bernays, I do plan on experiments later, really." But he was not an experimentalist. It was the detail of leaf anatomy that absorbed him—endless observation and lengthy, verbose description that wearied his committee members to desperation.

Years earlier, Reg and I spent Christmas in Costa Rica collecting grass-hoppers. We stayed for three weeks at La Selva Research Station in lowland rainforest, and for us it was the perfect Christmas—the two of us oblivious of anyone but each other, walking in the forest as hidden birds and crickets called, as lizards plopped onto broad, wet leaves, as jeweled butterflies darted in and out of sun spots, and howler monkeys threw nuts at us from high in the treetops. We were collecting grasshop-pers to study their jaws, holding hands as we walked along, until one of us spotted some camouflaged creature and pounced.

Mostly we were silent; Reg was a man of few words. I pondered again the nature of life, its exquisite complexity and immense diversity, and felt prickles of wonder at the ephemeral nature of each leaf or monkey, grasshopper or human, hearing—as poets over the centuries have—the litanies of passing beauty. Just as a lovely sunrise would become less precious if it lasted all day, so the shortness of life gives it a special value. The gold mariposa lily at Catalina State Park and the sulfur-colored paloverde blossoms in my garden in spring are lovelier for the short span of their lives. Each moment of life must be lived in full, because it will be extinguished with a powerful finality. Each day is one where I look at the natural world and am amazed—and have no recourse to thoughts of a hereafter.

Over beans and rice in the cafeteria each day, we exchanged findings with other biologists—Marla, who studied fruit-feeding birds' foraging tactics; bearded Geoff, studying reproductive behavior of poison arrow frogs; Linda Mary, with maps, trying to figure out moth migrations through Central America. At night, we worked in the lab—looking, measuring, recording. One pattern that emerged from our comparisons was that species of grasshoppers that specialized to feed on tough plants like grasses and palms had very large heads for the sizes of their bodies, while those that fed on softer leaves had heads of modest size. The reason was that to cut through tough leaves over and over required strength, big jaw-closing muscles, and, consequently, big heads. Furthermore, the jaws of grass specialists had great grinding surfaces with parallel ridges—elephant molars writ small, by over one thousand times.

From the grasshopper evolutionary tree, it became evident that the grass-feeding design of such jaws had evolved on at least seven different occasions. One can see this over millions of years as the groups diverged; seven branch tops involve species with these particular jaw types. This was exciting evidence for the importance of the adaptations and a great example of convergent evolution. Later, when I gave a talk at a meeting of vertebrate biologists in London, there were gasps of amazement as the picture of a single jaw of a grass-adapted grasshopper filled an enormous screen, displaying the pattern more familiar in the molar teeth of grazing mammals.

The study in Costa Rica made me interested in jaws properly. Insects have jaws that move in a horizontal plane (unless they are of the piercing-and-sucking kind, such as those of mosquitoes, or the licking kind, such as those of house flies), a fact that seems to fascinate children when I give talks in primary schools. Cool! Gross! Neat! Weird! They will exclaim. Or they laugh at the ridiculous notion. But when I get them to look at how the jaws fit together, snip bits of leaf, and grind grass blades, they become quite engrossed and begin to ask questions. Why are they sideways? Do they get worn out? Do they have baby teeth? How fast can they eat? Can they bite iron? And we go from there. I show them the simple jaws of caterpillars and the complex ones of grasshoppers, explaining that caterpillars are like overgrown embryos, while grasshoppers are more sophisticated, more like miniature adults. We discuss the pros and cons of different lifestyles, the fact that there are numerous solutions to the challenges insects have, and some of the children usually beg me for specimens to take home and watch so that they can see again how those jaws do their quaint sideways work and so that they can show their folks at home.

An interesting detail emerged from a study of a grass-feeding caterpillar called *Pseudaletia*. I took the eggs laid by one mother and divided them into groups. One lot of caterpillars was fed on tough grass, one on less-tough grass, another on soft grass, and a final lot on a very soft and nutritious synthetic diet made of crushed beans and extra vitamins set in a jelly. The tougher the leaves, the bigger the heads became, even though the bodies grew equally in the different treatments, and in dissecting the heads, it was clear that the jaw muscles were much bigger in those big heads. So, just as with people, muscle activity in caterpillars can increase muscle development and, in turn, influence the size and strength of skeletal parts—in this case, the hard exoskeleton of the head. Such variation caused by experience is often called phenotypic (as opposed to genetic) plasticity.

It was with this work that I had interested Frank and suggested he examine the caterpillars of *Phryganidia*, creatures that skeletonize very tough, old oak leaves—that is, they eat all the tissue between the network of veins, leaving behind a leaf skeleton. It seemed a potentially worthwhile topic, because the literature had focused on the importance of soft young leaves for insect herbivores. Young leaves are, after all, moister, softer, and richer in all the major nutrients. Why does a species neglect this resource to feast only on mature foliage? Frank never did find an answer. He was too polite to ever want to challenge the accepted belief that young leaves were invariably better for little herbivores.

My jaw studies also engaged the interest of Dan Janzen, a well-known ecologist who spends much of his time studying big moths in Central America. He sent me cases full of pickled caterpillars he had reared over the years. There were two families of moths represented—the hawk moth family and the atlas moth family. Their caterpillars grow very large—three to four inches long.

"Look at them," he summoned. "Find out something interesting. Let me know if you need more and I will send them." It was just like Dan to land me with something like this. He did it to others, too. He commandeered the efforts of biochemists to test his hypotheses about strategies of plant chemical defense and of biophysicists to study color and camouflage.

Email was in its infancy and, months later, in answer to my written queries about why he thought they were worthy of interest, he wrote back that the hawk moth larvae fed on young leaves of toxic plants, usually specializing on certain plant species. By contrast, the atlas moth larvae fed on a broader range of plants and enjoyed older, tougher leaves. There were consequences for biochemistry, for parasites and predators, for their ecology, for conservation even. Perhaps the two groups differed in how they processed the food for digestion.

I was busy with projects on insect learning at the time, but flattered that Dan thought my earlier work on jaws might have some application of interest to him. I put all the jars on a shelf in my lab and waited. Next year, I thought. But Dan didn't forget. He wrote offering me also the frass (fecal pellets) from their rearings, the cast cuticles, and yet more individuals.

Once he returned to the United States, he called me. "Liz, for god's sake, are you going to do something with those critters? If not, I can give them to someone else."

"Well," I replied, "I plan to get to them next week."

My communications with Dan were always brief. Although I think we are almost opposites in personality, we each knew instinctively where potential for new discoveries lay and where unexplored gold mines might turn up those nuggets of knowledge that have evolutionary implications. I hadn't planned to abandon the jars of caterpillars.

The results turned out to be more fun than I had imagined. Hawk moth larvae have complicated jaws for caterpillars. Each horizontally moving jaw has a tiny saw along the bottom, clusters of uneven teeth along the cutting edge, and uneven bumps on the inside surface. They are designed to close on soft leaf tissue and tear off small pieces that can be shredded as they are swallowed. By contrast, the jaws of atlas moth larvae are bulbous and simple, with sharp crescent-shaped cutting edges like nail clippers. They are designed to snip off curved particles of tough leaves that are then swallowed intact. The process can be watched, modeled, and the results checked by looking at stomach contents and frass.

Dan was fascinated by the contrast. "It must be something to do with different amounts of crushing required to deal with the differing leaf chemistries."

I argued with him. "It is much more likely to do with the mechanics of handling different materials. Clearly, a pair of snippers good for something hard or tough is no better than scissors on wet cloth if soft young leaves are to be tackled. By contrast, the grabbing jaws of hawk moth larvae would have no luck on hard, tough leaves."

I analyzed the data, wrote up the story, and sent him the manuscript, whereupon he altered the discussion to suit his ideas. I compromised in the revision, but gave my interpretation prominence, and he let it go. Watching the caterpillars in action gave credence to my ideas and fit, also, with another story I had been working on—the riskiness of different behaviors. With a group of students I had shown that, out in the wild, feeding is one hundred times more dangerous to insects than resting and that the smell, sound, and sight of feeding activity quickly attracts predators. Not only do the little herbivores need to handle food efficiently for the sake of extracting nutrients, it is critical that they feed efficiently for speed to avoid predation. Eat fast, rest at leisure. Reminds me a little bit of American restaurants, where people want to eat and be done with it, unlike in many European places, where the joy is to spend a whole evening eating and talking.

The big caterpillar project fascinated a few evolutionary biologists. It was more than just a comparison of two lifestyles. For example, among the hawk moth caterpillars was one species with jaws like those of atlas moth caterpillars. I queried Dan on his classification; "These guys are surely in the wrong group," I wrote. But he was right as usual. He really knows his pet group of insects.

He was back in the United States again and called me. "Yeah, pretty interesting. That species feeds on tough leaves that probably have a few tannins but no toxic alkaloids." The exception was particularly fascinating, because here was a species that had adopted atlas moth habits and utilized tough old leaves, and sure enough, evolution had selected for individuals that were better at handling those tough old leaves. Over time, the jaws had become perfected for the job and become generally similar to those of atlas moth caterpillars. More convergent evolution.

Another finding was that every species was unique. Each had some small feature that was not shared by any other. Each species could, in fact, be identified by the particular shapes or detailed architecture of its jaws. The differences could be chance alone, but I secretly believe that each is adapted to its particular host-leaf structure and that an extra bump or tooth may help to handle the peculiarity of its food plant, the crystals or hairs or cuticular spines or whatever. The power of natural selection is such a power.

So what would I say to my neighbor on the bus coming home from work when he asks me what I do all day in my laboratory? I don't think I would discuss jaws, unless I sensed that it would amuse. I think I would say that I study plant-insect interactions—how insects choose their foods, why they eat what they eat, and how this can sometimes help in understanding and managing pests. But actually, it is the details of comparison I love, for that is where ecological and evolutionary interpretation lies. The interpretation of details, the detection of convergent evolution and phenotypic plasticity, add another little piece to our understanding of how the living world works and evolves in nature, of how even we ourselves might be changing. And I shiver at the brevity of each life, each set of insect jaws, a brevity that carries the power of transient beauty.

Creosote Gold

Alone now, I examine a creosote bush in the desert by my Tucson home. Reg, my soul mate and partner, died here in springtime when all the yellow blooms gave us our last together-joy on a clear blue day. And then the paloverde petals blew away in the warm, dry wind, and the prickly pear flowers closed to a crinkly brown paper in the heat. Now, it is spring again, with all its yellow glory, a hot sun and blue sky, but it is a straggly creosote bush that fills my senses, its little shiny leaves and furry seedpods, its potent smell, and the memories of creosote days with Reg that crowd into my head.

My introduction to creosote, or *Larrea*, was at Boyd Deep Canyon in southern California. With Reg, I had traveled south from the University of California, Berkeley, to examine the grasshoppers that were reported to specialize on creosote bush, eating nothing but its sticky, resinous leaves. We were entomologists with a summer research project, getting away from the everyday tasks on campus. But most important, we were fascinated by the unusual, in this case two species of the grasshopper family Acrididae that confined their diets to a single plant type instead of the more common grasshopper habit of eating many different plant species. We stayed in the canyon together. The two of us always worked together, and being alone allowed us to be ourselves with our affection for one another.

First, we simply watched. There was *Bootettix*. It fed during the day on the leaves and hid among them, exhibiting a remarkable degree of camouflage. The base color of the insect was green, but the hue varied with that of the foliage. After rain, when the leaves were bright green, so were the grasshoppers, and in dry weather, when the leaves turned khaki or even brown, the grasshoppers were perfectly matched. Bright white spots on the insects blended with the reflective highlights on the resin-covered leaves, and brown triangles on them blended with brown triangles of the leaf bases. The insects were extremely difficult to distinguish among the clusters of creosote leaves, and later, when we needed to collect them, we beat the branches, holding a net below to catch them when they fell. Even looking at these insects, you knew they had to be specialists on creosote because they would have been too quickly found by predators had they ventured onto any other plant type.

The other grasshopper species was *Ligurotettix*, a gray insect that spends the daytime on the gray stems of creosote, again wonderfully camouflaged, with dark specks matching those on the stems. If you approach, it will swivel around to the far side of the stem. It took the two of us on either side of a bush to watch its limited daytime activity, and later on, it took the two of us to trick them into small collecting jars. Not surprisingly, this grasshopper feeds at night, creeping around the foliage where it would not be very well hidden in daylight hours.

The two strategies for feeding and hiding were an interesting contrast, and there was visual evidence that they were specialists in different ways on creosote, but we were there to do more than look. We wanted to understand what features of the plant made them stay and what behaviors were involved in their affiliation with this seemingly noxious plant that is distasteful to mammals.

We collected individuals of both species and ran experiments. *Bootettix* grasshoppers rarely leave their protective, leafy home. Their antennae respond to that creosote smell, and the taste buds on their feet and mouthparts respond to specific creosote bush resin chemicals, tastes that make them take a bite, tastes that are missing in other plants. An antioxidant compound termed nordihydroguaiaretic acid (NGDA) is present on the leaf surfaces and can account for 5–10 percent of the total weight of the leaves. This chemical particularly stimulates

Bootettix at these high concentrations. *Ligurotettix* likewise responds to the smell of creosote, and its odor receptors are highly sensitive to the particular vapors. They move between bushes and can find new ones by smell, the smell stimulating feeding.

Both grasshopper species have somewhat fewer taste and smell receptors than other grasshoppers, a reflection, we felt, of the simplicity of using a limited spectrum of special host plant chemicals for choosing their food instead of requiring input from a greater array of chemicals from hundreds of potential food plants among which choices would have to be made. But it is also true that, in these desert habitats, there are fewer plants to choose from, and that in the driest times, creosote foliage is the only food available to leaf feeders. Both species have adapted to use a reliable resource in a tough environment, each having a different strategy for becoming inconspicuous on the plant. As the desert slowly evolved, we thought, the insects would have gradually changed from being mixed feeders, like most of their grasshopper relatives, to being specialized on the one constantly available food.

Working together day and night in the desert, with our insatiable curiosity about plants and insects, we learned to appreciate the lives of *Bootettix* and *Ligurotettix*. Occasionally, *Bootettix* is very abundant, the bushes teeming with individuals that fly out when one approaches, but they usually circle and land on the other side of the same bush. It wouldn't pay to be off that bush, because they would stand out so strongly on the other gray bushes or the pale desert pan. When they are at low density, males tend to spread out among the bushes so that there is but one per bush, whose low song repels other invading males. Mostly, though, this species is unobtrusive, and being so totally camouflaged in the foliage, it is rarely noticed by people who are not particularly interested in insects. These are your low-key insects, good at hiding, able to eat at any time, with specialized mandibles having big opener muscles, able to handle their favorite leaves with the highest levels of resin that would make your average set of mandibles stick tightly shut.

On the other hand, if you walk in the desert in the heat of a summer day, you often hear click-click, click-click, click-click. The sound is part of the Sonoran, Mojave, and Chihuahuan deserts. It is the courtship song of male *Ligurotettix* grasshoppers, which are attracting females with their monotonous daylong clicking. In any patch of creosote bushes, the first male to reach adulthood looks about, tastes different bushes, and then selects the best one for his health and that of any females he attracts. It becomes his territory, a territory he will defend from rival males. Because he is first, he becomes aggressive and he clicks loudly. Late developers dare not sing so loudly and hesitate to come into his bush. Indeed, top clickers' vocal feats ensure that the latecomer lowers his voice. He is subordinate. He must make his principal home on an inferior bush. And even there, his voice will be less noisy.

From time to time, top clicker gets rivals in his bush who may sing quite loudly. Then he turns up the controls; he doesn't just click now, but clippers and clappers furiously, facing down the intruder, who at last backs off and retreats to a lesser bush. But it all takes time. Top clicker is kept busy defending his bush, keeping control of a territory where he expects to attract his mates.

The larger, silent females develop more slowly, and when they get there, they best like the noisy male. He lords over a harem of healthy females in his lovely bush, while lesser males make lesser sounds on lesser bushes and rarely get a mating. What is a timid little male to do? This is not an unusual dilemma in the world of animals. In the case of *Ligurotettix*, the timid male creeps quietly onto the best bush. While the top clicker is competing with his almost-as-noisy rivals, pushing them off with loud crackles, timid silent-guy rushes in and secretly mates with females standing by. If he is lucky, he may mate just as much as top clicker. Even if he is a small, dreary silent guy, he might have more offspring than vigorous, noisy top clicker. By sly speed, a weaker fellow might win and beat the brilliant, brutish showoff. Just as in humans.

In my Tucson garden, I have *Ligurotettix* clicking in the heat of summer, but we never found *Bootettix* here. I have many creosote bushes, though, and I examine the leaves, the flowers and fruits, the bare stems. I see those little caterpillars that also mimic leaves and a katydid, with the same kind of camouflage as *Bootettix*, and I see some galls made by flies. I pick a few leaves, crush them between my finger and thumb, and raise them to my nose. I breathe the smell and think of rain. I think of those days at Deep Canyon when I had a soul mate to work with, and we labored with a love of life as we watched grasshoppers day and night and dreamed up experiments as we lay in bed in an old trailer, surrounded by cages and nets, vials and magnifiers. In that tiny space, we set out feeding trials, extracted chemicals from leaves, tested orientation behaviors, ate frugal meals, and smelled creosote in the air.

I touch the sticky leaves and the hairy little fruits. The bushes are rangier here than the compact bushes in Deep Canyon and a little different from those in New Mexico, too, but they smell the same. In fact, those in the Chihuahuan Desert are diploid, with two sets of chromosomes in each cell, whereas in the Sonoran Desert, they are tetraploid, having four sets of chromosomes, and in the Mojave Desert, they are hexaploid, having six sets of chromosomes. The differences, though, are small, and the number of chromosomes apparently bears little on their chemistry.

All those past times of grasshopper study rush before my eyes when I touch the plant, and I relive the sense of sweaty heat, the potent plant smell, and the calm of togetherness. I don't remember our conversations. I remember the silences, the communication that floated noiselessly from one person to another, the sudden flashes of realization that we shared, and the collective joy of discovery. Are all the threads of thoughts that bound us still somewhere? Those days that seem so long ago are with me still. The unremembered talk is still embedded in my uncon-scious like the smell of creosote that is undetected from behind closed windows. The details of the experiments are no longer clear, but the warmth of doing them is still strong.

Creosote is one of the few plant species that keeps its leaves even in the searing heat of June, but then it is straggly and a khaki green, each little leaf coated with shining resin, each leaf base with a triangle of sticky brown. I have read that creosote leaves may contain so much resin that it makes up 50 percent of their weight. When Reg and I first saw *Bootettix*, it was in the heat of summer, and we were astonished at how the little green grasshopper of the books was leaf-colored khaki. Reg quickly said, "Pretty good case of phenotypic plasticity." He sounded casual, but I knew we felt the same wonder at such perfected camouflage. We didn't need to speak to know our togetherness.

Now, a few small seedpods remain from the yellow flowering that I saw after the last spring rain. The little hanging balls coated with soft white hairs are themselves a decoration on this desert shrub, and their contained seeds a rich resource for many desert animals. More than all the species of cactus, I find creosote bush to be at the heart of the southwest deserts. After rain, the bright green leaves and golden blooms are part of spring, and then, as the desert dries once more, the leaves they keep that turn almost brown are summer.

For Reg and me, creosote always had a special importance, with its intense aroma filling the air as moisture arrived, our tension rising with the excitement of rain as we sat under a porch in our Tucson home, his hand on mine, each of us remembering Deep Canyon though we rarely spoke of it. For all who live here in this desert, few natural phenomena are as magical as rain, when the smell of creosote bush permeates everything. Reg and I experienced that together for twenty years, and now I must take it in alone. One day recently, I was walking in the desert with a girl of about eight, and she picked a sprig of creosote and held it to my nose. "Smell the rain," she said. She had never experienced rain without the smell of creosote. I smiled, for I have often crushed a little creosote and sniffed it, thinking of summer storms that give the desert its life, remembering the thrill of rain and the days of waiting for it, remembering all the creosote years with Reg.

A few years after we worked in Deep Canyon, we accepted invitations to come to Arizona, to work at a very different campus in the desert habitat we both had grown to love. And here, in a house surrounded by creosote-filled desert, we built on those first desert works and demonstrated that, generally, being a food specialist allows insect herbivores greater protection from natural enemies, because without the complexity of choice, their senses are freer to maintain vigilance.

Creosote bushes have lived in North America for perhaps a million years, after birds brought seeds from South America. Each plant can live a hundred years or more and can produce side shoots that live on when the principal plant dies so that, eventually, a circle of plants replaces the original. Mostly, though, the plants are evenly spaced in the desert, competition for water preventing new plants from germinating in the intervening spaces. And so, over hundreds of miles of dry terrain, where rainfall averages as little as three inches per year, a pattern of creosote plants gives the North American deserts a character that is distinctive and different from other deserts of the world. Unlike the plants, the two grasshoppers that specialize on creosote in North America have their phylogenetic origin here in North America and must have evolved their specialist traits after the introduction of creosote. It was a story waiting to be studied, as Reg so often remarked. We planned to use DNA to show that theirs was a case of extremely rapid evolution, and I dream of working on the project still, but it is just a dream without him.

Here in Tucson, creosote plants are dotted around the city. They fill vacant lots, line certain streets, and are sold in native-plant nurseries. On our four acres, they were scattered everywhere, mixed with cactus, paloverde, and bur sage. Reg died having a view of it all from the window and, in those last weeks, we often remarked, "Hasn't it all been wonderful!" Now I love the rain in particular, because when it comes, I am in touch with my past both at Deep Canyon and here in Arizona as the smell comes in through my windows, and I run outside to breathe deeply and remember the satisfying times of finding things out together with a mate, our joy of sharing work, our fascination with grasshoppers, our constant questioning of why there were such different diet breadths. I look back on a life in science and feel that Reg and I were among the lucky ones. We shed tiny rays of light on small biological questions, but the payoff in our lives was in gold.

Reflections on Mortality

When I was a college student in my first biology class, the professor said, "Mortality is everything. Take fleas—unless most of them died, the world would be knee deep in them in a year." It was the stuff of freshman biology lectures, perhaps exaggerated, certainly memorable. Years later, after becoming an entomologist myself working on plant-feeding insects, including agricultural pests, the mortality issue returned vividly as I became aware of how the feeding habits in those small herbivores result from a web of effects, all of them related to death.

One summer twenty years after that freshman class, I was in Costa Rica on a field trip with Reg, and we walked along a path where howler monkeys threw seedpods, keel-billed toucans called, and all the rainforest bristled with life. It seemed a good place to tackle my recent interest in mortality of herbivorous insects.

"Let's try and quantify the predation risk to plant-feeding species here," I said.

"That's a tall order," Reg replied with a frown. "I mean, how can you actually quantify such a thing?"

I was frustrated; he always seemed to see the difficulties. I decided I had to do something. Next day, I picked a spot of vegetation on the edge of a clearing and randomly chose a cubic-yard space that made a virtual observation plot. In that cubic space, I watched for every small animal continuously for an hour. There were plant chewers such as caterpillars, beetles, and grasshoppers, along with all manner of plant-sucking bugs. The enemies of the plant eaters included jumping spiders, assassin bugs, marauding ants, and lots of different predatory wasps. I found that numbers of enemies exceeded those of plant feeders by nearly two to one. I hurried back to show Reg.

"Well, one data point doesn't say much," Reg retorted.

I had to agree that one observation could have been chance. So, each day for several weeks, I spent an hour watching a cubic yard of forest edge. The time passed quickly, as I kept seeing new species passing through the plot or peeping out from behind leaves while I listened to the life of the forest. The persistent, highly potent risk of mortality was clear, and that did not even include the predators that come out only at night or the big predators such as birds, lizards, and frogs. I was excited to be able to tell Reg that the data were there.

Over the years, I watched a lot of predation events outdoors. Of 800 caterpillars on a California oak tree, I recorded 793 kills over a nine-day period, just by paper wasps. Argentine ants foraging at night took all fifty-three eggs laid by cabbage butterflies in a weedy patch of nasturtium. Predators killed 99 percent of hornworm caterpillars feeding on their native plants in Arizona before they were even a few days old. There were extraordinarily high levels of predation wherever I looked. Massive mortality.

Vulnerability to predation was greatest when the plant-eating insects were feeding. During one summer in California, my students and I observed small caterpillars on French broom plants, recording predation on the caterpillars during feeding events. We put chairs out beside the plants for the all-day work, and, after a couple of weeks, we had watched a total of 296 caterpillars. In these long observations, in sun or fog, little predatory bugs did the killing, and we found that the risk of predation during feeding was one hundred times the risk of predation when caterpillars were resting. A similar pattern was found with wasps attacking different caterpillars.

Why should feeding be so dangerous? we asked. Head and jaw movements, chewing sounds, smells from newly damaged leaves all seemed possible reasons. Whatever the signals the predators detect, the caterpillar must eat its meal as fast as possible and get back to its shelter or its stationary, cryptic pose. It would pay to have a good chewing apparatus. In fact, each species turns out to be fine-tuned to the physical characteristics of its particular host-plant leaves so that eating can be speedy. Take a look at the mandibles of a caterpillar feeding on old, tough tree leaves and you see structures much like a pair of sharp metal cutters; look instead at those of a caterpillar feeding on soft, fleshy leaves and you will see a multitude of pointed teeth and serrations for tearing the tissue. The chewing apparatus is different for every type of caterpillar-leaf combination. Such minutia are highly significant in the lives of insects, but easy to overlook because insects are so small. When I gave talks, though, the message really took hold because the photos were highly magnified.

A majority of plant-feeding insects are restricted to certain plant species, a fact that has always interested biologists. After all, being able to feed on many plants should be a safeguard against running out of food and should also allow plant eaters to live in lots of different places. Conventional thinking was that plants defended themselves from insect damage by producing toxic chemicals, so to feed on any plant, an insect must become especially good at dealing with the particular chemicals of that plant, sometimes even using the poisons for their own protection. Monarch butterfly caterpillars, for example, specialize on milkweeds and have perfected their ability to deal with milkweed poisons. By contrast, the less common generalists, such as gypsy moth caterpillars, would be jacks-of-all-trades and masters of none. The idea was attractively, deceptively simple.

I said to Reg, "I think the emphasis on plant toxins as a reason for specialization might be unbalanced."

"Of course," he replied. "Plants have other reasons for their unusual chemicals—think of all those diseases. It's just that theorists love to simplify."

We decided to examine the roles of plant chemicals in the lives of plant-eating insects more thoroughly. Do the chemicals actually taste bad? Are they really toxic?

I offered insects edible wheat-flour communion wafers (purchased from monasteries and blessed by the priest) that were either impregnated or not with different chemicals. It turned out that chemicals from all the plants they didn't eat were distasteful to the insects. Certainly such chemicals restrict what each insect eats. However, a bad taste does not necessarily signal that the chemicals are poisonous. Perhaps the chemicals provide signals for an insect not to eat the plant because the plant is unsuitable for other reasons, such as being a bad place to avoid particular predators. Are the bad-tasting chemicals poisonous? How to get insects to eat them?

Sometime later, shy, Cornell-trained Barbara joined the lab and was enthusiastic from the start. "How about making tiny coated pills?" she suggested. My idea had been to develop tiny stomach tubes for dosing, but her idea was better.

So, together we visited a pill-making plant in Palo Alto, where we learned how to make microcapsules—in this case, tiny crystals of test chemicals coated with sugar or plant wax so that insects could eat the stuff without tasting it, just as our bitter-tasting drugs are coated to allow us to swallow them without tasting them. All manner of plant-feeding insects got doses of these now-edible materials painted on their normal food. We measured how well they survived and grew, and we found that most of the bad-tasting chemicals were completely harmless to all the beetles, caterpillars, and grasshoppers we tested. This strongly suggested that restricted diets were not about avoidance of poisons.

Barbara had gone, but I presented the case at a lab meeting. "Camouflage comes to mind: if you look like plant A, it will be safer to stay with plant A than to leave for plant B with better nutrients. Look at all the spectacular photos of caterpillars that resemble pine needles, oak catkins, flower stems, and at the variety of shades of green matching the green of the host plant. I have to believe that staying on, and feeding on, a particular plant really might be related to escaping predation."

Reg, of course, was the first to criticize. "Those benefits are more likely to evolve after a species acquired its narrow diet, rather than causing the narrow diet in the first place."

"Well, yes," I had to agree, "but it is part of the puzzle." We needed to know other things. For example, are specialists generally less vulnerable to predation than generalists, or are the spectacular examples of camouflage just rare cases?

My first test was with common paper wasps. These predators hunt caterpillars with which to feed their young, and some of them had constructed big nests with hundreds of individuals inside my big, old greenhouse in Berkeley—a ready-made opportunity to test the idea that two of my students were keen to take part in.

Jeremy, Michelle, and I made big cages, each two meters cubed, in which we placed a wide variety of specially grown plants. We introduced twenty healthy caterpillars of a generalist species and twenty healthy caterpillars of a specialist species to a test cage a day before the tests. One of the plant species was eaten by both caterpillar species, so the environment was naturalistic for any pair of species being tested. A day later, when the insects were settled in, we took the top off the cage, and the predatory wasps soon found their way in.

"Here she comes," cried Michelle as the first wasp walked on a plant and dipped her antennae, touching the leaf and anything that might be on it. All three of us gazed at our first forager. Suddenly, the wasp encountered a caterpillar, grabbed it round the neck with her jaws, and squeezed until the hapless creature's green blood spurted all over the leaf. Then she ripped open the caterpillar, pulled out its digestive tract, and rolled the remains into a ball before taking to the air. She would feed it to the larval grubs that live in the cells of the intricate house she helped to make, where the queen wasp laid her eggs. It is theater that anyone who spends a lot of time outside may see, a performance of great skill. We were spellbound.

"Wow, watching death is so fascinating," said Michelle.

"I can't wait to see the end results, though," said Jeremy. And that was my sentiment, too.

Michelle, Jeremy, and I watched many hours on many days for months, comparing numerous replicates of twenty-six species pairs. Sometimes a wasp discovered the generalist first, or a wasp walked over the specialist, apparently not even noticing it. Some specialists were rejected after wasps touched them. In some cases, the wasp took a long time killing the specialist and ignored the same kind of caterpillar on subsequent visits. On each occasion, we waited until wasps took

50 percent of one species and we compared the mortality rate on generalist versus specialist. The three of us pored over the results later and, though we were not surprised, the mortality was always much greater for the generalist of the pair.

"Hey," said Michelle, "one up for predation!"

Mary, another student, said, "Jeez, I wish I had been involved. Can I give it a try with predatory ants?" She did her own series of experiments with Argentine ants, which were nesting all around the Berkeley area. The pattern was similar. In all experiments, predators took the generalists more often than the specialists. But the question remained: Even if specialists benefit from restricting their diets, did specialization begin for some other reason?

More discussion ensued in the lab on how to proceed. I loved these weekly meetings where everyone had a go with ideas and suggestions. Les was all for the conventional approach of highlighting plant chemistry as the reason for specialization, but the others liked my hypothesis that predators mattered more.

John said, "How about comparing predation on two closely related species that look the same but have different feeding habits?"

"That might help," Reg responded. "At least it reduces some of the complications."

Heather had been reading about closely related moths with different diets. "I want to work on them," she said, eyes shining. She chose the common pest *Heliothis virescens* as the generalist that feeds on hundreds of different plants. It is called tobacco budworm, tomato fruitworm, and other names depending on which crop it attacks. It also feeds on ground cherry. For comparison, she chose an almost identical species, *Heliothis subflexa*, which feeds only on ground cherry.

Heather grew the plants, reared the insects, and carried out all the experiments. With her baby-doll looks and high-pitched voice, she surprised us all with her hard work. She found that the generalist, *H. virescens*, grew best on sunflower and tomato and not so well on ground cherry. The specialist, *H. subflexa*, fed and grew only on ground cherry. The wasps found the generalist fastest on sunflower and least well on ground cherry. The wasps also found the generalist more than the specialist on ground cherry.

The results gave a picture in which one could imagine how the specialist, *H. subflexa*, might have evolved from the generalist, *H. virescens*. In a situation where a predator such as the paper wasp was abundant, the best thing to do would be to specialize on ground cherry, since it provides more protection even to a generalist. Perhaps the specialist evolved in a place with a lot of wasps. Heather and I were excited. The role of predators in influencing the affiliation of insect herbivores and their host plants was getting attention.

Then there is vigilance. It is well known that selective attention is critical in the lives of all animals, and there are mechanisms to focus attention on the most important factors at any one time. A generalist (at anything) has more choices, must make more decisions, and may need more time and attention for tasks than a specialist with habitual neural pathways. In nature, such differences can be expected to translate into different levels of vigilance, and consequently, different levels of competence in escaping predation.

Attention to danger is reduced when any animal feeds, so why not with insects? Seeing the dangers firsthand, watching the killings, it seemed that vigilance had to be of singular importance. I thought that focusing on a single plant type and having simplified signals relating to acceptance or rejection of food would allow improved attentiveness to danger, so I studied the taste cells of caterpillars and grasshoppers and found enhanced responsiveness to a few chemicals in specialists, implying that they might be able to focus more easily on a food plant, allowing improved vigilance.

The new task was to test whether decision-making was more efficient among specialists than among generalists. The first system involved generalist grasshoppers. I had already found they were easily trained, so the idea was to train them to be either specialists or generalists. This strategy had the advantage of testing different patterns within a species, rather than having the additional complications of comparing different species. I trained individuals of *Schistocerca americana* to be specialists on one of six single food types or generalists on a mixture of the same six food types, and then, after a month, the grasshoppers were given their tests.

On test day, each isolated individual had a choice of the six different food types. Those trained as specialists went faster and more directly to their training food, were less hesitant to start and complete a meal, and stayed with the one food to complete the meal faster. They were, in other words, more efficient than the generalists and foraged for a shorter time to eat the same amount—a good thing for reducing mortality. Even Reg liked this.

One day at a meeting in another lab, I met Dan, a postdoc who was working on a type of greenfly that has specialist and generalist populations in different parts of the United States. He was excited to work with me using his aphids, and we sat long hours in a greenhouse together. We had a great time in those days, discussing philosophy and religion during the inevitable times we had to wait for the greenfly to do something. We learned that the specialists discovered the best food faster, hesitated for a shorter time when the food was reached, and found the plant sap with their proboscises sooner after settling on the food.

"Boy, I believe you now," Dan said. "I need to look at the beetle data from my PhD to see if this is something I should discuss or even work on."

Further evidence came from whiteflies, moths, and beetles, and, in all cases, the specialists were more efficient in their feeding behavior; they made faster decisions and foraged for shorter times. Inevitably, they would be able to better monitor other factors in the environment, such as the presence of predators and parasites.

Nowadays, the early emphasis on insect-plant chemical coevolution has to give way to a more comprehensive story involving at least three factors: the plant, the herbivore, and a variety of enemies of the herbivore—the so-called tritrophic approach. The risk of death is so high for the herbivores, it is a major selective force in their host affiliation. The importance of selective attention and vigilance, so well studied in humans and other vertebrates, is no less important in insects. It is reason enough to expect specialists to predominate, allowing later evolution of refinements such as exceptional camouflage.

I was enchanted when Reg finally said, "You know, I think you are right. Of course, I always thought the emphasis on plant defenses was overdone. You're the one to upset the applecart!"

Those observations and experiments, which started off with such excitement and proceeded in so many directions, have led to the conclusion that mortality due to predators drives much of the evolutionary processes involved in the specialized affiliation of insects with their host plants. Rates of killing are so high that any small factor that can reduce risk will be selected for. Vigilance, and all that is needed to enhance vigilance, will be paramount, including factors such as eating quickly, making decisions rapidly, using chemical signals for ease of host choice, and reducing the number of tastes that compete in the brain for attention.

Although the small size of insects deters many researchers, I think that examining the details of their behavior can significantly increase one's general biological understanding. I think of Darwin's perspective on evolution and on complexity: the few survivors in each generation who are best at avoiding death pass those skills on to their offspring. Such a simple truth, so wonderfully explicated 160 years ago. It's all been a great surf ride for me, and I agree now with that first professor in college: mortality is everything. Yet the characteristics insects develop to avoid mortality are multiple—vigilance, neural economy, learning ability, feeding-apparatus differentiation, shape and coloration of camouflage, behavior, chemical tolerance. Mortality drives change, death in the service of evolution. And what an exciting ride it has been to spend my life solving problems related to insect mortality and how it matters in the lifestyles of all those herbivores, discovering how we can harness that knowledge to regulate agricultural pest populations.

Prickly Pear Persuasion

Some love it as a plant they know in its proper habitat, a native plant in deserts of North and South America; or as a crop for the cultivation of cochineal insects for their red dye; or simply as an exotic potted plant that is carefully tended in climates unsuitable for its growth. In many countries around the world, though, it is hated as a weed, a plant that can grow uncontrollably fast in places where it has been imported.

Prickly pear plants are made up of pads joined together, which are actually flattened stems. The only leaves are small, short-lived fingers that cover the new, expanding pads in spring. Scattered over the mature pads are areoles from which long spines and fine brown bristles, or glochids, grow. Each areole is also a growing point from which a new pad, flower, or root can develop. Any pad broken off and allowed to lie on the ground will put down roots when it rains and soon become a new plant. In this way, prickly pear can spread fast.

I sat one November day in Montrose Canyon near Tucson. It had been hot and dry for months. The pads of prickly pear were thin and wrinkly—diamond shapes between spines, rows of ridges like ripples on a sandbar. They looked old, for surely, wrinkled means old. But they were mostly not old. With winter rain, they would swell and fatten, become smooth with no sign of wrinkles. Not many living things rejuvenate themselves like cactus pads, though they cannot regain the tiny, fat pink-green fingers that were their leaves. I thought how, inside those wrinkled pads, there was still moisture, waiting, storing life under those green skins, and marveled at the adaptations to the droughts that occur in southern Arizona before and after the short-lived summer rains.

Although there are hundreds of species of prickly pear, varying from miniature plants to trees one hundred feet tall, the common ones are wide, spreading shrubs about four feet high, with pale, gray-green rounded pads about nine inches long. The Engelmann prickly pear (*Opuntia engelmannii*) is one of the commonest, with its long white spines and rusty-colored glochids. Across the sun-bright Sonoran Desert of Arizona, the classic scene is patches of round, clumped pads in their faded watercolor green between giant saguaros and paloverde trees. Looking at a steep, rocky slope from a distance, one sees the smooth, hard surfaces of prickly pear reflecting sunlight at different angles, those in shade enhancing the green of their sunlit neighbors by contrast. No other place in the world resembles the deserts of the American Southwest.

It rains. The muted greens of cactus plants seem brighter without sun, and drops roll off the prickly pear pads to darken the desert pan. Moisture causes the strange fragrance of creosote to pervade the atmosphere. Cottontail rabbits have taken cover and no birds sing. Rain is such an event in the desert. Through the melancholy air, I watch the gentle fall of contentment. This winter rain will make a colorful spring. The pads of prickly pear are swelling, the wrinkles of drought ironing out as the plants' masses of fine roots absorb the merest suggestion of water, those roots that hold together the thin layer of desert soil.

From every window of my house here in the Sonoran Desert, I see Engelmann prickly pear, though patches of different cacti, bur sage, and creosote bush are intermingled, along with a wide variety of less common desert plants. The diversity is remarkable to me after having known the Sahara, where the few plant species are scattered and rare. And with this rain, the seed bank in the soil will sprout dozens more species that will wait for a little more warmth to grow up and flower.

How much the prickly pear grows in the spring and how many flowers develop depends on how much rain we get this winter. I think of April, when the young, soft, bright green nopales swell and are eaten as a vegetable by many, especially Native Americans. I think of bright, shiny yellow cups that sit on the plants like botanical jewels. I think of the cactus bees that swim about in the abundant pollen and then leave, with their pollen baskets full and their body hairs laden with tiny, yellow balls. And the flowers that open on a second day, which are peach colored and less sunny. After flowering, when the desert is again very dry, fertilized flowers produce small green fruits that wait out hot June and July, becoming wrinkled as the pads do. Eventually, summer storms bring the rain that allows the pads to regain their fleshy appearance and fruits swell wine red, providing a rich crop of sweetness for birds and insects. This plum-sized fruit, called tuna, is also another human food.

I remember my father telling me how, as a young man farming with his brother in Queensland in the 1920s, he had to spend so much time every week cutting down prickly pear. "Ugh, that menace," he would say, standing up and putting his hands on his hips whenever we looked at pictures from that time. "Ruined our farm. Ruined my life." And my mother would join in: "Wretched stuff. You would go out in the morning and, as soon as you turned your back, there was another acre of it."

Prickly pear started to worry people in Australia the late 1800s, but it was years before the first Prickly-pear Destruction Act was passed— when landowners were obliged to destroy the new pest, and inspectors came to check on them. By 1900, the new weed was out of control. A ten-thousand-pound reward was offered for the discovery of an effective control method, but it was never collected. In the cities, politicians had no idea about the magnitude of the problem, but one of them did take twenty-seven fellow politicians on a train tour to see firsthand the impact prickly pear was having on rural areas. I can find no reference to the details of this tour, but I imagine the wonder that those men felt looking at mile upon mile of solid cactus. Photographs from the time show nothing but dense prickly pear for as far as the eye could see. The trip must have been effective, anyway, because soon after, the Prickly-pear

Destruction Commission was formed and given wide powers to deal with the problem.

However, the situation seemed quite hopeless, and, by 1925, prickly pear infested twenty-five million hectares in New South Wales and Queensland—an area almost as big as Texas. It spread at the rate of a half million hectares a year, and nobody could stop its progress. Tremendous effort went into mechanical and chemical treatment programs, but the weed could not be contained. It was during this time that my father was kept busy with mechanical control, every spare hour spent cutting the stuff down. He and his brother worked with hatchets and machetes every afternoon and all weekend slashing the prickly pear, but of course, it was able to grow again from the broken pads. In our family, everyone had to hate the stuff as father did until the day he died.

All kinds of methods were tried to control the problem that had become desperate, destroying ranches and pushing farmers off their land. In 1926, the Queensland Prickly Pear Land Commission report stated that the amount of poison sold in Queensland that year would treat 9,450,000 tons of prickly pear. The poison chemicals included 31,100 twenty-pound tins of arsenic. Some 335,000 birds, mostly emus, were also killed in various ways, based on the mistaken belief that birds were responsible for spreading the cactus.

Besides activity in the two affected states, the Commonwealth government was also busy. In 1919, its Prickly Pear Board was working on possible biological control agents. Several insect species were imported from the Americas, and a laboratory breeding and quarantine station was established in Brisbane. Eventually, a moth, *Cactoblastis cactorum*, with brightly colored larvae became the savior. The caterpillars proved successful at eating out the fleshy insides of the cactus pads. Researchers tested the insects in labs to ensure they would not move onto other plant species before finally releasing them in 1926.

Then came mass rearing of *Cactoblastis*, with two to three billion eggs distributed in affected areas during the late 1920s; a fleet of seven trucks and one hundred men distributed packed eggs across the state. The aptly named insect was spectacularly successful in destroying the weed. By 1932, the *Cactoblastis* larvae had caused the general collapse and destruction of most of the original, thick stands of prickly pear. Many millions of hectares of previously infested land were made available to settlers. Abandoned farms were reclaimed and brought back into production.

The amazing spread of prickly pear in eastern Australia was considered to be one of the botanical wonders of the world, and its virtual destruction by caterpillars is still regarded as the world's most spectacular example of successful biological control of weeds. Dr. Alan Dodd, the entomologist in charge of the program, echoed the feelings of many when he wrote that "the most optimistic scientific opinions could not have foreseen the extent and completeness of the destruction. The spectacle of mile after mile of heavy [prickly] pear growth collapsing *en masse* and disappearing in the short space of a few years did not appear to fall within the bounds of possibility."

As a Queenslander, I feel proud of the research done back then and the amazing effectiveness of the biological control program. As a student, I visited Dr. Dodd's laboratories in Brisbane, though he had retired by then, and the lab was involved in other projects as part of the Commonwealth Scientific and Industrial Research Organisation (CSIRO). I suspect that my father's tales of his youth and the wonderful story of a weed controlled by a caterpillar influenced the way I slipped into entomology as a career, into studies of herbivorous insects, and even into the study of weeds and their insect parasites.

A movie was produced called *The Conquest of the Prickly Pear* to record the success of the biological control program, making *Cactoblastis* a star of the big screen. But the most lasting memorial to the success of the program was built in what had been the center of prickly pear infestation in Queensland. The residents of the district built a hall and, in gratitude to their entomological benefactor, named it the Cactoblastis Memorial Hall. It is near my birthplace in Chinchilla, Queensland.

Some years ago, I visited the area to find the hall still there, a simple wooden structure built on stilts with plaques nearby explaining the whole story, and I have a photo of myself sitting on the steps. I found plastic *Cactoblastis* caterpillars for sale in Chinchilla, many of them made into little brooches, and I wondered who would actually buy one, let alone wear one. Perhaps those whose families benefited from the wonderful control program feel some affection for the famous caterpillar. In Dalby, southeast of Chinchilla, a cairn was erected in Marble Street in 1965 to record the indebtedness of the people of Queensland to *Cactoblastis cactorum*. I saw the rich agricultural land of the Darling Downs, a landscape of rolling hills, pastures, vegetables, cotton, wheat, barley, and sorghum mixed with stretches of park-like bush having creeks and herds of cattle. Here and there, windmills pump water from the Great Artesian Basin, and scattered remnants of the past—rusty old woolsheds and farm equipment—lurk among the silvery brigalow trees. Once, before I was born, this land was nothing but dense prickly pear jungle.

During my visit, I saw occasional prickly pear bushes in dry spots under eucalypt trees. Certain refuges such as these are resistant to serious attack, but they harbor small populations of the *Cactoblastis* caterpillars, thus ensuring that the moths do not become extinct and can continue to keep the weed in check if it begins to spread. I looked everywhere for a caterpillar, but they are rare now; their population is just enough to take off again if ever the problem returned.

With the story of a devastating prickly pear from my youth and a family that scorned any interest in cactus, I never could take an interest in cactus gardens or the cacti I saw in plant nurseries. In biology classes, I was always taught about cacti as weeds, not just in Australia, but also in Africa, India, Hawaii, the Caribbean, and southern Europe. When I first visited California, I discovered that prickly pear was even a problem on Catalina Island off the southern coast. On Ascension Island in the South Atlantic, where there had been another successful eradication of prickly pear, there was even a postage stamp printed featuring *Cactoblastis* moths and caterpillars. I, too, truly admired those insects that ate the cactus.

It wasn't until I lived in Arizona and saw prickly pear in its natural habitat that my feelings about the cactus changed. When the Engelmann prickly pear flowers bejewel the pale gray-green expanses of cactus, I have to love this plant. The butter-yellow cups open as the sun rises high, then fold as the sun goes down until the next day. When I see birds feeding on the dark red "pears" that swell after summer rain, I am thrilled at the richness of this desert. I sit and watch woodpeckers tearing into a new fruit, white-winged doves following, cactus wrens, thrashers, finches, and mockingbirds finishing them off. I notice the seed-filled red droppings on my patio, and I know the birds are having their summer feast. I glimpse butterflies in the sunlight, moths and bats in moonlight, all feasting on the sweet harvest of fruits that have been opened by birds or squirrels.

But the prickly pear plants are not out of control in Arizona. Here they are kept in check by a variety of animals. I have seen jackrabbits and cottontail rabbits feed on the pads. They deftly manage to avoid the spiny masses, leaving behind little bouquets of spines around a chewed plant, or they spit them out. Pack rats chew the flesh between spines, letting scraps fall, and they take the uneaten spiny bits to put on top of the mounds of debris that mark their nests. Javelina feed voraciously on the pads, swallowing everything including the big spines. When I walk through the desert, I see few prickly pear plants without some kind of damage. Some are masses of half-eaten pads and others are eaten down to a couple of nubs near the ground.

There are insects, too. A long-horned black beetle, the *Moneilema*, with some antennal segments bright white, feeds on the pads, making holes or notches in them; its larvae feed inside, burrowing through the flesh. When I see a healthy pad with a neat hole or a notch that gives it a heart shape, I know *Moneilema* has been there. The prickly pear and the beetle adorn the poster for the Entomology Department at the University of Arizona, a poster commissioned by me from the Tucson artist Paul Mirocha.

Then there are the cochineal insects, dark red and covered with white wax, which insert their fine mouthparts into the flesh and suck out the juice. When I squash a mass of white wax on prickly pear, the red juice oozes out and stains my fingers. In the heat of summer, when the plants are stressed, the little red insects with their white coating reproduce quickly, and whole pads may be covered in the white, fluffy wax. Tunics and gowns used by the royalty of Aztec, Inca, and other early American cultures were dyed with cochineal. Its value was greater than gold for them. After the Spanish brought back the dried cochineal insects to Europe, the dye was used for many items, not least of which were British military redcoats and Roman Catholic cardinals' robes. I remember as a child seeing the small bottle of red food coloring in my mother's cupboard, with its "cochineal red" label. Mother used it for making the pink layers of sponge cake that would be sandwiched together with strawberry jam, and she once let me dye one of my handkerchiefs by boiling it in a solution of cochineal. Today there is once again a cochineal industry, a return to natural dyes after one hundred years of synthetic aniline food colorings.

I know many less conspicuous creatures living among the prickly pears and using the resource as well. There are large, brown, flat-footed bugs that attack the pads and fruit, stinkbugs that suck the developing fruit, and cicadas that suck the juice from the roots, causing whole plants to collapse and rot. When I walk in the desert, no prickly pear is without signs of biological stresses—mammal and insect attack, fungus, or bacterial diseases. There is no *Cactoblastis* here, but other natural enemies keep the balance and prevent prickly pear takeover. Its value is clear, and its role in the balance of nature reminds me that I live where so much of the wilderness is just as it has been for millennia. Even in suburban Tucson, there are patches of pristine Sonoran Desert filled with prickly pear that provides food for so many animals, including people. This plant is one I have grown to like and enjoy.

I realize now, as I look at all the cacti living in a community of native desert plants in harmony with native herbivores, that the critical issue of balance is what makes the difference. Lack of balance leads living things to become pests, both plants and animals; the naturally balanced habitat is something to admire and enjoy and stimulates questions about the workings of nature. Weed control may have sparked my interest in solving man-made problems, but undefiled nature has been a stimulus not just for my entomological career, but also for an aesthetic appreciation of all plants and animals, of their color, diversity, cohabitation, ecological interactions, and evolution—an appreciation of life itself. Improbably, I have come to love prickly pear in this Sonoran Desert.

Sierra Interlude

Silence at last. We hadn't made it to the overlook in time for sunset, but a suggestion of red glowed over a series of angular mountaintops. A picture-book crescent moon highlighted the darkness overtaking the sky—all seen through silhouettes of lightning-damaged pines. It felt so fine to be back in California, in the High Sierra, in the cool, dry air of eight thousand feet. And the silence was good.

Freddie was the first to speak. "Big storms up here—not a tree without a high, dead branch. Look at that one, shaped like a tuning fork."

Molly laughed. "I thought you were going to say something else." She didn't elaborate and none of us especially wanted to hear, knowing it was likely to be another dildo joke.

As star brilliance replaced red dusk, the group began to talk again, but in almost whispers. Finally, Julie's voice, "Must get back to work, I suppose," reminded us of why we all were here.

"Okay," said Freddie. "Gotta go, gotta go."

We piled into "Taylor," the truck, and Freddie slid it slowly down the rough track with the headlights off, allowing us to enjoy the moonlit scrub, the starry night.

"Hey, Freddie, you forgot the lights," Molly giggled.

He sighed slightly and gave us the bright view of the path we were bumping along, turning the mysterious night into a cozy truck full of field biologists going back to camp.

I had joined the camp a week earlier, keen to see how the field-crew members were doing their experiments, keen to examine the butterflies, and hopeful that I could use the system they were working in to test my own theories on insect foraging and the problems of being a herbivore. This was to be my first extended field trip after the death of Reg, so I was also glad for the distraction of working with a different group of enthusiasts led by Freddie. He was a friend of over twenty years, and I had read many of his research papers about the Edith's checkerspot butterfly (*Euphydryas editha*). He had worked in Californian field sites for thirty-five years and at this particular one near Kings Canyon for twenty. His knowledge of the butterfly was prodigious, and his long-term study of genetic change in behavior over time—evolution in action—was unprecedented.

This year there was Freddie and three students: Julie, Molly, and Dave—four highly contrasted individuals, but it would take time to find out about them properly. For my visit, I was honored with a sleeping space in "Lambert," the old trailer that Freddie towed up the mountains every year. The others slept in four of the six tents in camp, though Julie's tent was out of sight among the lodgepole pines and ceanothus bushes.

Lambert turned out to be a mixed blessing. There was the bunk, certainly, and with screening, an absence of mosquitoes. The cooking facility for camp—a pentane stove—was also convenient for early-morning cups of tea while everyone else still slept. The broken-down fridge contained a useful library. However, Lambert was also the place for night work at the small table; electricity from solar panels powered lights along with an electric balance for weighing larvae and pupae. Gossip and work continued until midnight, except that Julie's work usually went on until the small hours, long after I was undressed and in my sleeping bag on the darker side of the cramped space.

Between me and the workstation was the stove, usually covered with dirty saucepans and frying pans; the sticky blackened floor, with food coolers brought in so as not to tempt bears; and the shelves, with plastic cups of caterpillars and pupae and all kinds of paraphernalia needed for the fieldwork. At the foot of my bed was a heap of Freddie's clothes and papers, Dave's shoes, bags full of sticks, markers, wire, and other unidentified objects. The space above, including over my bed, was hung with cages of butterflies. The small sink was full of variously aged corn, onions, potatoes, and oranges, there being no running water. The small toilet was for emergencies, as emptying it would be sometime in the distant future.

It was delightful to get out early and alone into the fresh, still air, to see the first rays of sun through the trees, to hear the woodpeckers and watch the mountain chickadees, to walk among the great sweeps of phlox, collinsia, paintbrush, and monkey flower, and then across the meadow full of shooting stars and daisies. The view south included snowy peaks. I was the first to get the spade and toilet roll from their spot under the largest ponderosa pine and walk to one of the sandy areas with dense bushes. I was the first to get a splash of water on my face from the carboy outside Lambert, avoiding the forlorn bowl of dirty dishes under the spigot. I was the first to forage for bread or fruit or cereal from among the various boxes, cupboards, and coolers.

By seven o'clock, Julie was checking small dishes of eggs or caterpillars, putting pupae in or out of the sun, and examining butterflies in cages hanging on the long clotheslines strung between two Douglas firs. She took breakfast from her own store of health foods in her isolated truck and sat down at a small table in the early sun to sort newly hatched larvae into new cups. Slim and boyish, with dark curly hair, long legs, a dimpled smile, and gentle eyes, she bent over a cup and peered into it with a hand lens. She had an immediate appeal.

"How's it going, Julie?'

"Fine, thanks, Liz. Busy day." She smiled. And then she would have some technique to discuss: "Do you think I need to worry about the fate of eggs in the field, or should I just put out newly hatched larvae? I would like to have the females lay the eggs on the plant, but so many eggs are taken by predators, I don't think I can get enough data on larvae if I start with eggs."

On another occasion she began, "I don't want to handle the larvae. Should I let them just climb out of the epi tube themselves?" And so we would talk, turning over the details of each experiment, trying to establish the best compromise for getting the data she needed for her question. She knew the literature on evolutionary ecology and all the relevant studies of plant-feeding insects, including my own. She was curious about many details but had big questions.

Julie studied speciation: How do two species arise from one? Evolution of new species is thought to come about when populations become physically isolated, allowing them to develop different characteristics in the different places (allopatric speciation). Among plant-feeding insects, it appears that speciation can occur without geographic separation of populations (sympatric speciation), and Freddie's team had established, using molecular techniques, that genetically distinct populations of his beloved "spotties" used different host plants—collinsia or *Pedicularis* or plantain or paintbrush. Females of each population had behaviors adapted to the physical details of the specific host plant—details of landing, curling the abdomen to lay the eggs, or arrangement of the eggs on the plant. This, then, could be a starting point for the evolution of new species.

Julie's plan was to cross individuals that used different hosts and to compare performance of their offspring with performance of those reared from pure strains. If a butterfly species is splitting into two, hybrids between the two are typically weaker in some way, the best offspring coming from same-kind matings. Julie's careful experiments would, for the first time, test whether there was evidence of the beginning of this kind of speciation occurring in a single geographical area.

Julie had a goal with a specific focus. She worked with speed and precision. She worked long hours. And she loved the outdoor life, the physicality of camping, though her eyes misted over when she talked of her lover, a woman working in Tanzania that summer.

After Julie, Dave was usually the next to emerge from his tent. He stretched. "Ah, fabulous day, what luck." Then he stood on his hands by the picnic table and jumped up over a chair before racing off down the slope for his morning run. Julie and I laughed. But he was only away about fifteen minutes. He approached us, firing his comments:

"Hi Julie, hi Liz, great day, good day, good morning! Say, you girls had breakfast yet? Must just run to my outcrop site." And he picked up his backpack ready with vials to collect plants with spottie eggs on them and was gone.

Dave was examining movement of butterflies within and between populations at different subsites near Kings Canyon. It was important to know if changes in host preference could be due to movement of butterflies with different preferences or could be the result of genetic changes in the butterflies resident at a site. He and Freddie did a lot of preference testing—a time-consuming study in butterfly behavior.

A female butterfly in a waiting cage is gently picked up by the wings and placed on a potential host. If she is interested, and if the smell is right, she taps with her front feet on the leaf surface and lowers her antennae, a complex tasting procedure. A highly acceptable plant causes her to curl her abdomen after just two taps with her feet. A somewhat less acceptable one requires more thought; she taps four or five times before curling. Marginal hosts require repeated tapping and may not result in curling at all. Unacceptable plants never elicit curling. She abuts her curled abdomen upon a leaf, and one by one, small, green eggs emerge and lightly stick to the leaf and to each other. In about twenty minutes, there may be fifty eggs in a batch. The method varies with the host. On *Pedicularis*, she likes to nestle down in the crevices at the base of the plant and push the eggs downwards. On collinsia, she may walk delicately on an upper leaf and put the eggs in lines along the petiole. "Pedic" and "colly" specialists are not the same, however. Pedic females occasionally accept colly, but have trouble with it; they fall to the ground looking for the elusive crevices in this spindly plant. Colly specialists walk around on the prostrate pedic plant with abdomens waving upwards, apparently seeking the more erect stems of colly.

Dave and Freddie worked together most days. Freddie loved to reiterate the wonders of spotties. "Just look at that, Dave. Come over here, Liz. Look at this colly kid on pedic. Guess what pop she comes from, eh? What an indevicive girl!" Freddie's eccentric habit of inverting syllables or even words, or altering their pronunciation, was one of his rather endearing characteristics. This female butterfly was being tested on colly and didn't curl. She was then tested on pedic and curled after just two taps, but was not allowed to lay eggs, as this would have altered her readiness to lay eggs at all. She was retested on colly to ensure the difference was not just due to the order of plants offered. She rejected it again after eight taps. She was definitely a pedic specialist. On to the next butterfly.

Freddie emerged from his tent in his crumpled clothes of the previous day. His blue denim shirt was missing a button where it was tightest over his small paunch, and the white skin of his behind showed through a hole in his gray baggy pants. His nearly bald head was sunburnt and the remaining wispy gray hair straggled down his neck and round his ears.

"Good morning, Freddie," I called from Julie's table. "Want a cup of tea?"

"Arghhhh."

"You okay, Freddie?"

"Arghhhh."

"Yes, then?"

"Diarrhea. No tea." He strutted slowly to the spade and disappeared over the hill.

Meanwhile, Dave was back. "Hi, guys, anyone want an egg, over easy in butter?" But he knew he was the only taker. He wiped out the dirty frying pan and searched the coolers. Ice had turned to water in all of them and a smell of very old broccoli filled the air.

"So who's going to get ice, eh? Pongs here." He found the eggs and butter and disappeared into Lambert singing. In a few minutes, he appeared at the door.

"Anyone seen the thingo for turning eggies?"

"Try Freddie's tent," I suggested. "I think he had it for scooping up some spilt seeds in there."

Dave found it eventually, with a dozen rubber bands around the blade and a small note tucked into them. "Remember to collect more males from Tamarack," he read before skipping back to Lambert. Freddie, returning, repeated it: "Remember to collect more males from Tamarack. That's what you need, isn't it, Julie? And don't we want some more capertillers from the R5 site?"

Most mornings, Freddie forgot breakfast as he checked on the various livestock in camp. There were trays of pupae waiting to emerge into butterflies on a table under a shade cloth stretched between four young lodgepole pines. He peered into the cups through their transparent lids. "Aha, a Tamarack female and two Pine Tree males."

Next he checked the potted plants. There were collinsia seedlings grown in the natural soil, large flowered and small flowered varieties.

"Liz, look at this. See the senescking ones?" I examined the little plants that were senescing, or beginning to die. "Some go yellow and some go red. The collinsia butterflies from stone ridge prefer the red-senescking ones and grow better on them, but they lay the eggs long before the plant gets to that stage, and the pedic butterflies that will accept collinsia don't discrinimate. Fantastic, eh?"

Next, he checked the exotic plants he had obtained from nurseries, which contained certain chemicals called iridoid glycosides. These were the host-plant chemicals that spottie butterflies particularly liked. Were there some species that all butterflies would accept? It would have been useful for Julie, perhaps. He proceeded to his cups of larvae being reared. He had two types: small ones that would go into diapause, the resting stage, in a week or so and stay that way until the following spring; and larger ones that had hatched the previous summer, were in diapause through last winter, and would soon become adults. I offered to feed them and got detailed instructions. "Just *so* much fresh food. Too much and it gets too wet in there, and the moisture on the sides of the cup is an impemident to lomocotion. On the other hand, too little and it's not enough to last a day."

Freddie sat on a chair beside the box and talked. He told and retold his spottie findings, his stories of evolution and behavior. He was proud of his long-term studies and all the details of the evolutionary history he had uncovered. He reminded me of a comment I'd made to him years ago, long before I'd seen the butterflies and discovered how docile they really were. "You said," he began, laughing, "'I have to say your methods don't look at all feasible, except that they produce such reasonable results.'"

"Well, yes, it has to be seen to be really believed!" I replied. And I remembered when he first told stories of holding butterflies by the wings in order to test their plant preferences. I was not the only one who thought the method preposterous—most of us in the entomological world were more than conscious of how stress altered insect behavior, and nothing, usually, was worse than handling for stressing an individual and making it behave abnormally.

He sighed and removed his very dirty feet from his sandals, the heels worn through around the edges. "I have had such trouble with grants over the years because people didn't believe how wonderfully anemable spotties really are. But now I think they do."

"I am a believer now," I assured him, and he leaned over to peck me on the cheek as he chuckled into his scruffy beard. He had always been a kind friend, but it was nice to have him say so this way. Among our colleagues, I was perhaps the most critical of his rough-and-ready methods, and he was ultrasensitive to criticism. I put off the discussion I had been waiting for—my theories on the importance of natural enemies. We had diverged in our views on plants and insects over the last fifteen years. Freddie was certain that the plant was the principal reason for specialization, whereas I was certain that predators and parasites had more to do with it. I would wait a few days before discussing the experiment I wanted to set up.

As Freddie talked and I listened beside the box of cups, Julie went on sorting eggs and larvae at the outdoor table, and Dave ate his hearty breakfast in Lambert's doorway. Slowly, out crawled Molly from her tent, shaking her brown hair with its bleached and bright pink strands. Yawning and stretching in front of her tent, she spoke slowly. "I doan know how you manage on so little sleep, I gotta have eight hours at least."

Freddie smiled. He liked this youngest of the students, though she was still without any research focus. "Rise and shine, Molly. Go and get some breakfast, and then you can help with butterfly feeding."

Molly was obese but seemingly not concerned about it. She returned to her tent and came out with three chocolate bars. "Okay, I better get all the cages into the sun then?" she said.

"Yeah, and bring over the stored feeding pads so I can lick them and taste them for concetration—they are in Lambert somewhere. Oh dear, they are too concetrated. Let me just add a bit of water."

Molly took down about fifty of the lightweight pentagonal butterfly cages and placed them on the ground in the sun. By the time she had finished her chocolate, there was fluttering of wings and a general waking up of butterflies. She sat on one of the stray chairs, turned to face the sun, and took a cage onto her lap. She unzipped the side nearest and slid in a Petri dish containing a foam pad soaked in artificial nectar. The butterflies rested or fluttered on the sunny side of the cage, and she picked up each one by its closed wings and placed it on the pad. Obligingly, as their feet touched the sugar, out came the proboscis and feeding began. Females can overeat, so each of them was allowed no more than two minutes before being chased off the pad. They don't find the sugar on their own in the cages, so this controls these gourmands' appetites.

"Freddie, I have one here that won't feed. What am I gonna do with her?"

"I'll come over. She will need to have her proboscis unrolled and placed on the pad. Here, just use this needle and poke it into the roll of her proboscis, then gently pull it away, see?"

"Food and sex is all these things do, eh? And they can't do anything in cages. Look at this old male. Do you think he is any good for crosses, any sex in him?"

"Old doesn't mean no good at the romantics, Molly," Freddie chuckled.

"Ha ha, no offense meant, Freddie. By the way, when you going down into Fresno? I wouldn't mind doing some shopping. When you went down Monday, I asked you to bring cake, and all you brought was nuts. Oh, and we need more vials from Smart & Final. By the way, Freddie, you snore. I distinctly heard you last night."

"Now, my spouse has never said I snore. She said she couldn't live with a man who snores—that's her semintent."

"Well, you do. My boyfriend snores a little bit. But I just kick him gently and he stops—well, for a while anyways. Unless we get busy."

As the morning warmed up and everyone concentrated on their jobs, conversation dropped off, though at intervals Freddie started again on a spottie story. Dave would have had jokes, but he was at another field site. The fierce California sun climbed and, under hats, heads were damp.

"How about lunch?" I ventured.

"Sure," was the answer from all.

I decided to find materials for a salad and get away from the individual foraging syndrome. In one cooler was a pack of mixed greens. In another, the tomatoes I had brought with me. In Lambert, I found an onion and a pepper, some walnuts, lemon, and olive oil. Dave was back just in time. He remembered his last trip to Fresno and a lovely girl he saw in Kinko's, where he went to plug in his computer and get email.

"Gorgeous—two taps curl, for sure!"

"Say, Liz," Freddie said, "what about a trip to Tamarack, eh? It's only ten miles as the fly crows, but we have to go down the mountain, then north and up to 7,500 feet again, okay?"

"Sure, whatever." I was happy to go along and see everything.

In the previous ten years, I had steadily worked on different systems to test a new idea. Perhaps the restriction of most plant-feeding insect species to narrow ranges of plants was related to the avoidance of predation. After all, an insect could have a more sophisticated camouflage on a single plant type than the generalized green or brown that was only crudely camouflaged on any particular plant. Specific cases were famous—the pine looper caterpillar is exactly like a pine needle, and the grasshopper is a wonderful replica of grass blade. The notion was unpopular when I began testing the hypothesis; plant chemistry and plant defenses were fashionable, together with the belief that narrow diets were about adaptation to specific plant chemistries. The examples of perfect camouflage were seen as unusual, or at best, the end result of eons of time on particular hosts after adaptation to the plant chemistry.

Time passed and experiments proved that specialists escaped predation more than generalists, but here, in the California mountains, with spotties having preferences for one, two, or several different host plants, I could perhaps manage a completely naturalistic study.

Freddie was keen but felt he knew enough to be sure I was wrong. Anyone working on the system he considered his was inevitably deemed a bit of an amateur. "I know the persolanity of 'em Liz, and they ain't like your grasshopper guys."

Julie was intrigued and, being younger, seemed more open. "Be great if it worked."

Next day, Freddie and I went to Tamarack. With nets and vials and new butterfly cages, coolers of food and bottles of water, we wound our way to the ridge. We had many stops: Stone Creek, where, eight years previous, a fire came through and the quality of host plants changed so that colly insects began to prefer plantain, which had been introduced from Europe; Glenwood Flats, where Freddie first found paintbrush-feeding caterpillars; and eventually Shadow Gap, where we began to see flying spotties. It took me a while to get the technique of catching them: Females fly close to the ground searching for host plants and nectar plants; males fly higher and faster and more erratically, seeking out females and avoiding other males. The differences require different netting methods. We mainly needed males for Julie, but then Freddie decided it would be nice to get females as well, to check that their host preferences were the same as last year.

We stopped for our picnic lunch, jumping up to net the occasional passing spottie, and as we rested, I mentioned my plans again. I would have to coax butterflies with preferences for one or several hosts to lay eggs on the various plant species and then study the fate of eggs and caterpillars in detail. Time and endless patience would be needed. Long hours of watching tiny creatures would be essential. But the desire to get an answer bugged me so much, I was compelled to do whatever it took. I had learned from Freddie and his team how to handle individual butterflies, but it would be a long job to get fifty individual females to lay eggs on appropriate plants, count the tiny eggs with the aid of a lens, mark the plants, and record everything relevant.

Back at camp at the end of the day, I had one last discussion with the team about my plans. We decided Pine Tree Meadow was the best spot. It was an easy walk and close to where Julie had experiments running.

Freddie was skeptical. "Liz, you'll be dead lucky if you get good data. And you know, I have looked at natural enemies. Nothing in it." But he was happy to help with the egg-laying experiments and impart the details of his spottie knowledge.

Julie knew it would be a long job. "Good luck," she smiled, as Freddie fussed over exactly how I should hold each female butterfly and get her to do the job.

Molly gaped, "You really gonna spend that much time?"

"Good on ya," was all Dave could say, laughing at his attempt to be Australian.

For many days, I lay on my stomach in Pine Tree Meadow, wearing an OptiVISOR (magnifiers on a headband) and hat, long pants, and long sleeves. I recorded predators, deaths, losses. I ignored the night hours, returning to my patch at daybreak to see what had disappeared in my absence. Certainly, eggs and baby larvae disappeared. There was a suggestion that the offspring of more picky butterflies survived in greater numbers, and my excitement grew. Bets were cast in the camp at night.

"Good on ya, Aussie. Go for it," Dave encouraged.

"It's all over the shop each year, I tell you, and, for sure, there will be just a confuzzed data set." Freddie expressed his usual pessimism with a newcomer to his own special system.

Julie was hopeful. "Aw, Freddie, you haven't seen the trends."

"Increludous I will be if she finds a pattern."

I rose earlier each day, made tea, and rushed to my sites.

Molly brought me snacks. "Don't know how you can stand it. When do you pee an' that?"

It was nine days after I had started my observations. I hurried into my clothes as my tea brewed, examined my notebook, collected my OptiVISOR and vials. I felt that this experiment was to be a definitive test of my idea in a well-researched, totally natural system. Would the results support the growing interest in the role of higher trophic levels in host affiliation of plant-feeding insects? I would surprise Freddie. The timing was critical. The season was coming to an end, and I needed to use every hour I could muster in this last, enormous effort. The next week would provide the answer—yes or no.

Arriving at the first site at daybreak, I found that all the plants had been grazed to the ground. At the second, many were gone and critical labels were knocked over. And the third was a similar scene of confusion. I was aghast. I never did discover what animal had done the deed, but it was the sudden end of my experiment. Such are the vagaries of fieldwork. Such are the reasons why students take years to complete degrees in ecology. The group all said, "Too bad," with knowing looks, and Freddie laughed. "See!" I knew they felt for me, really, but each had experienced frustrations of one sort or another and knew that such is the nature of field research.

Though my disappointment was profound, I said, "Too bad," and laughed with the others.

Few outside biology know the enjoyment of fieldwork in a team of real enthusiasts, and few understand the resilience that must be part of being a field ecologist. Disappointments abound, but there is always hope for the next day, the next season, the next new discovery, the unexpected. And, at the end of the day, there is the friendly bonding among a group of colleagues who are passionate about their investigations.

Thirteen years have passed since my foray into the Sierra Nevada with eccentric Freddie and his gang. My memories are full of congeniality in the midst of mess and apparent chaos, of hard work and fun while surrounded by conifers and meadows clothed in alpine flowers, where myriads of butterflies went about their lives—butterflies and caterpillars that were subjected to the vagaries of life in nature and the selection pressures that change the proportions of particular genes in their populations. I found out nothing concerning the questions that forever rumble in my head, yet came home quite renewed. Maybe I will get the answers using spotties another time.

Cups and Nostalgia

As I sit by the side of a vegetable garden in Patagonia, Arizona, I see rows of tall, clear plastic cups on the raised beds. Each cup has a half-inch hole closed with screen in the upturned base, and each has a strip of fine wire netting suspended near the hole. There is moisture condensed on the inside walls of the cups, which cover bare soil and scraps of straw. Bean seedlings are yet to emerge. The cups have been my solution to a problem—birds snatching the plants before they can expand their first new leaves. In a few days, bean seedlings will emerge beneath the cups, and the fat cotyledons will push through the soil. Between them, the first tripartite leaves will unfold and expand. The magic of photosynthesis will begin.

Here, with my plastic cups in a garden, I remember. The cups have been out of sight in a cupboard for years, their earlier uses forgotten, and it is suddenly strange now to see them again, this time in a new role. They were not purchased and altered for use in a vegetable garden.

Each morning and evening, I survey the rows of cups, look through the plastic, lift a few of them just to see if there is yet any sign of disturbance in the soil. I sit lazily in the first light of day or at sunset, as towhees, sparrows, and cardinals sing in the mesquite trees around the garden and a mockingbird does his variety show on top of a utility pole. Years back, I saw grasshoppers on the wire netting near the tops of the cups, where a lamp above provided warmth. I spent so many hours modifying cups as individual grasshopper houses, so many hours and days and weeks observing the insects in a warm, dark room to find answers to questions. Decades of work. The grasshoppers basked up near the overhead lamp most of the time, but descended to the bottom every so often to eat the food provided there.

I had theories to test, ideas to investigate. These cheap, simple cups are the reminders of my career in biology. It was with such cups that Reg and I along with our student Steve continuously monitored individual locust behaviors for five consecutive days and nights, and so discovered that they had twenty-minute rhythms of activity, including feeding. It was in such cups that we discovered individuality among locusts: they had very different personalities in the ways they walked, fed, and rested. One (number nine) always rested on its back with legs in the air; another always had very long meals and intermeal intervals. Yet they all ate the same approximate amounts of the wheat grass presented to them and they all grew well. Number nine, though, is the individual I remember best because we thought it was sick. In fact, it was not sick, only different. I tell students about number nine to impress on them that, like all other animals, insects are individualistic.

It was in such cups that we tested chemicals for their effects on the feeding behavior of grasshoppers. Did mixtures of antifeedants have an additive effect? Or might they be synergistic? Did individuals become accustomed to them, causing the antifeedants to lose their potency, or did the chemicals always remain active? Were the antifeedants detected with antennae, with palps, or after biting? It was in such cups that we evaluated different plant species for their ability to support the growth and development of grasshoppers. In other experiments, we offered not their normal leafy diet, but synthetic food with precise quantities of essential nutrients for studies of nutrition; it was served in tiny glass dishes that could be weighed at intervals to measure amounts eaten. So many experiments using grasshoppers imprisoned in all those upturned cups.

Grasshoppers are good at learning, and I exploited their ability in order to raise individuals with different feeding habits, each insect in its own plastic cup. In one experiment, there were two main treatments. One involved six dishes, each with a different added odor. The other involved six dishes, all with the same odor. Grasshoppers learned that food came in six flavors or one flavor, and became either generalists (mixing their variety of foods) or specialists (learning that one flavor meant food.) The training period lasted a month, and then I tested my hypothesis that generalists take longer to make decisions than specialists do.

Movement in the wash below interrupts my grasshopper memories. I look through the wire fence at three white-tailed Coues deer that are making their way along a rough trail there, two does and one fawn. They stop once in a while and take halfhearted bites at the dead grasses. There has been no rain through winter and spring, and everything is dry and brown. The fawn raises a back leg and scratches its neck with a foot. I stand for a better look, and they run off up the slope on the other side, white tails raised like flags. And then I am back with the cups. Bean seeds swelling below, insects in cages from the past—nostalgia for the old excitement of testing ideas, of new discovery, and of arguments and camaraderie with colleagues and students. The work was an all-consuming endeavor, and I hardly noticed time passing.

On the big test day after the month of training, I presented each grasshopper with six dishes. This time, though, all individuals got variety— six different flavors, including one flavored dish that the specialists knew from experience to be the only food. I predicted that the grasshoppers that had become specialists would also feed more quickly and efficiently than those reared as generalists. By close observation of their every move, I monitored decision times and feeding rates. Sure enough, the specialists were faster.

I taught children to make such cups so they could observe grasshoppers feed and molt and think about the problems such animals had in their lives. It was always a hit, and the children would take the cups home to try watching all kinds of small creatures as well as grasshoppers. Sometimes they came to me later begging for more cups or more wire screen.

The very same cups used for my previous observations are the cups I look at now. I weigh the fun of those heady days engaged in scientific ideas against the pensive quietude of my life today. I look along the rows of cups on the soil and feel lucky for the life I have led, lucky for the days of cups put to work as insect houses. And I am still fascinated by how even grasshoppers take time to choose food when there is a choice to be made.

Actually, it is nothing new among humans or other animals that the more items there are to choose from, the longer it takes to make a choice. In our culture, having a choice is considered valuable and important; the more choice of any commodity, the better. Yet there is plenty of evidence that too much choice can be a problem, because making the right decision becomes a preoccupation and, in some cases, completely paralyzing. In a supermarket, a wide variety of items even leads some people to choose nothing, whereas a limited selection frees them to buy the item they actually want. The demand on our mental ability to make a decision among choices shows in trivial ways, too. It takes longer at a party to select an item from a tray of diverse finger foods than it does to select from a tray with just two foods. It takes time to decide what to wear, unless one has only a single set of clothes.

Even before a choice can be made in any particular situation, the available options have to be detected. No animal, including a person, can take in everything about a whole scene at once. Instead, attention to parts of the scene changes over time. In general, conspicuous items are detected first. When items stand out strongly, for example, with a bright color or unusual shape, we notice them quickly so that fast and accurate judgments can be made, in a similar way to how we decide when the choices are few. Selective attention to subsets of inputs from the senses is normal for our ordinary behavior.

Yet, my study with grasshoppers was not just idle curiosity about such issues. The hypotheses related to a bigger question: Is feeding efficiency a significant reason for the common characteristic of food specialization in plant-feeding insects? At the time, perceived wisdom attributed specialization of herbivorous insects to plant chemistry: those that are specialists should be better at dealing with particularities of plant chemistry than generalists that are jacks-of-all-trades. Supporters of this conventional approach were skeptical of my theory and wondered, did it really matter that generalists dither more in deciding what to eat? That they interrupt their meals more? Take longer to eat a certain amount of food? Out in the natural world, are generalists really hampered by their hesitancy?

For humans, it usually matters little, but for these small creatures, it is very important. One may ask how important a 20 percent increase in time taken to eat a meal could possibly be, but it is not the dithering or indecision itself that makes a difference. The repercussions are critical, because generalists, in processing so much information, have reduced vigilance with respect to risks and, as a result, fall prey to a multitude of predators and parasites. Before the study in cups, I had spent time over years observing predators taking insect prey in the field. The principal cause of death for numerous species of herbivorous insects, including grasshoppers, was predation, at times amounting to almost 100 percent mortality. Most predation events take place when the prey is engaged in foraging activities. Feeding is incredibly dangerous for insects, therefore they have an intense need to feed quickly and efficiently so that heightened vigilance can be resumed and death avoided.

My various findings were part of a big jigsaw puzzle, the pieces of which were choice, risk, vigilance, and the benefits of a restricted diet. In the natural world, a plant-feeding insect that is a specialist must still find, detect, and then select its host among a multitude of plants. So they have smell and taste organs that are hypersensitive to the key chemical characteristics of their hosts. The particular plants that they will eat or lay eggs on are thereby made conspicuous in the sea of plant chemicals that fill the air. My intelligent grasshoppers reared on one main chemical learned to use that chemical in their foraging behavior, and it turns out that caterpillars also get hooked on a particular set of plant chemicals, making it easier for them to locate and select a similar plant. Many flies, beetles, moths, and butterflies have evolved special sensitivity to unique chemicals that indicate their host plants. For cabbage specialists, the odor of mustard oils in the plant is particularly intense; for monarch caterpillars, the taste of cardenolides in their milkweed hosts is unusually strong. Such strong signals allow the insects to selectively focus on feeding without distraction, so as to complete their meals in the fastest possible time.

Finding satisfying answers to unsolved problems has given me intense pleasure and is often why scientists become so absorbed in their work. Upturning beliefs held with fervor is also fun and another stimulus to work at experiments. It is fascinating for me, as well, to contemplate the general idea of selective attention. How do we focus on subsets of incoming information, and in what different ways do individuals manage to pay attention to the things that matter most? I have an ability to focus long enough to obtain the scientific information described here, while those with attention deficit problems may be similar to the generalists in my study. Their behavior is less efficient, so their lack of perseverance doesn't lend itself to success at certain jobs. On the other hand, they may benefit from having observational skills that allow them to perceive a broader spectrum of phenomena, making them more quickly aware of possible problems or bonanzas. For instance, I'm absorbed in watching the car ahead of me, while my partner sees the cars on either side, the ones behind, the plane or helicopter above, and the oncoming hurricane.

As I sit in the garden with the birds and deer around, I gaze at the little cups. In reality, they were a small part of my entomological life, yet they engender intense nostalgia. I reflect on the meaning of that mixture of longing, love, and melancholy and all the days of working with a soul mate. Neurobiologists find that dramatic nostalgia is associated with one of the many types of serotonin receptors in the brain, suggesting a genuine function. Perhaps, though, it is simply a by-product of our awareness of death; psychologists put it that nostalgia solidifies and augments identity, quieting our fears of the abyss, but perhaps it is to "keep the wolf of insignificance from the door," as Saul Bellow suggests in *Mr. Sammler's Planet*.

So, I reflect with contentment on the small part I played in the bumpy progress of knowledge and concede the possibility that, just as most of the players are, I will be forgotten as the years pass. We are tiny points of light, like a mass of glowworms in a cave, each living briefly and passing on, but wonderful at the time. I am glad to have kept the cups for the memories of work done with such pleasure and for the indulgent but enjoyable nostalgia that increases with the years of my life. When I am gone, these cups will remain somewhere, probably with a life-span of more than a hundred years—trash with no significance for anyone, while birds continue to sing in the mesquite trees.

Acknowledgments

I thank my mother, who first advocated for my love of insects and flowers and who provided support and encouragement throughout my childhood. I am deeply indebted to math teacher Joy McCallum, biology teacher Hazel Gray, and Professor Patricia Marks, who believed in me and whose teaching ensured that I won a scholarship to university. I thank the many postdocs and students and colleagues round the world who were integral to my development as an entomologist, but above all my gratitude is to Reg Chapman, who trained me as a scientist and gave me thirty-seven years of love and companionship. When I turned to writing, I was fortunate to have encouragement and guidance of faculty in the creative writing program at the University of Arizona, especially Richard Shelton and Alison Deming. I thank my wife Linda Hitchcock for her support and patience as I wrote, and Karen Pickell of Raised Voice Press for her help in significantly improving this book.

CPSIA information can be obtained
at www.ICGtesting.com
Printed in the USA
LVHW050900160919
631186LV00004B/230/P